Table of Contents

Glossary of Key Terms and Definitions

Aerosol: A disperse system of particles suspended in a gas (which, in the context of this book, is air).

Aerosol (or particle) concentration: The amount of particulate matter or aerosol per unit volume of air. Usually this is expressed in terms of mass per unit volume of air (e.g., mg/m^3), but may also be expressed in particle number per unit volume of air, or particle surface area per unit volume of air. It may also be expressed for an individual chemical or biological subfraction (e.g., mg of respirable crystalline silica dust/m^3).

Inhalable particulate matter (IPM): The fraction of particles, defined in terms of a probability as a function of particle aerodynamic diameter, which is aspirated through the nose and/or mouth during breathing. This is equivalent to the aspiration efficiency of the human head (*see Sampler aspiration efficiency below.*)

Inhalability: *See Inhalable particulate matter.*

Occupational exposure limit (or OEL): A suitably time-weighted concentration of a given relevant aerosol fraction below which, according to certain criteria and qualifications, a person may be exposed for a prolonged period without risk of ill-health. *See also Threshold Limit Value (TLV®).*

Particle: An entity of matter which, in the context of this book, refers to a large number of molecules of the material in question sufficient in number for the particle to be defined as being either liquid or solid.

Particulate: Strictly this is an adjective that qualifies something as having the nature of a particle (e.g., particulate matter). However, in the context of this book, it is sometimes used as a noun (i.e., referring to a particle).

Particle aerodynamic diameter (d_{ae}): This is the diameter of a spherical particle of density 10^3 kg/m^3 (i.e., a droplet of pure water) which has the same falling speed in air at atmospheric pressure as the particle in question. It is strongly influential in how particles are transported in the air, are inhaled, and are deposited in the lung. By the same token, it is influential in the collection of particles by sampling instruments.

Respirable particulate matter (RPM): The fraction of inhaled particles, defined in terms of a probability as a function of particle aerodynamic diameter, which passes down to the alveolar – or gas exchange – region of the lung.

Sampler aspiration efficiency: The efficiency with which particles are extracted from the ambient air and enter through the orifice(s) of a sampling instrument. Usually this may be expressed in terms of the ratio of the concentration of particles of a certain size entering the sampler and the concentration of particles of the same size in the ambient air.

Sampler (or sampling) efficiency: The efficiency with which the particles of interest (e.g., in a given particle size fraction) are collected by a sampling instrument. Usually this may be expressed in terms of the ratio of the concentration of particles of a certain size collected by the sampler and the concentration of particles of the same size in the ambient air.

Sampler or performance: *See Sampler (or sampling) efficiency.*

Thoracic particulate matter (TPM): The fraction of inhaled particles, defined in terms of a probability as a function of particle aerodynamic diameter, which passes into the lung below the larynx

Threshold Limit Value (TLV®): An occupational exposure limit (see OEL) defined by the American Conference of Governmental Industrial Hygienists (ACGIH®). It is the time-weighted average (TWA) concentration to which "nearly all workers may be repeatedly exposed, day after day, without adverse effect." Other bodies use similar definitions.

Preface

Occupational (or industrial) hygiene is widely defined as "......*the anticipation, recognition, evaluation, and control of hazards in the workplace environment.....*" Within this definition is the assumption that the health and well-being of the individual worker should not be impaired during the pursuit of his or her livelihood. Since large proportions of populations worldwide spend significant parts of their lives at work, the subject is of great relevance to society. Driven by the resultant consequences to the social and economic health of nations, occupational hygiene therefore remains a high priority area for scientific study. Aerosol science has always been an important part because airborne particles that gain access to the human body by inhalation have presented some of the most prominent and intractable workplace hazards.

The key issue in occupational hygiene is *exposure*. It may be defined as the intensity, time-averaged in some appropriate way, of the agent of interest at the relevant interface between the environment and the biological system representing the worker. Here, for an aerosol comprising an ensemble of solid or liquid particles (e.g., dust, mist, fume, etc.), the intensity might be the airborne concentration (say, in mg/m^3), and the appropriate part of the biological system would be the region of the respiratory tract where the particles first comes into contact with the exposed subject. So, for example, for dust exposure in relation to silicosis, a disease of the alveolar region of the lung, the exposure of interest is the concentration of inhaled particles of crystalline silica that are small enough to aerodynamically penetrate down to and deposit in that region. We know from aerosol science that such penetration is a strong function of particle size. This therefore leads directly to the concept of particle size-selection as a basis for exposure assessment and, in turn, *occupational exposure limits* (OELs). This in turn leads to the concept of aerosol measurement, specifically *sampling*, relating to the act of removing a sample of particles from a known volume of air so that it may be analyzed to provide the mass concentration. The long history of such aerosol measurement in the occupational hygiene context was reviewed in a recent paper by Walton and Vincent (1998).

It was as early as 1913 that McCrea noted that particles in lungs of miners observed *post mortem* were restricted to the range from 1 to 7 μm in diameter, and so identified the need to selectively measure the finer particles as an appropriate index of exposure for certain types of dust-related lung disease. This appears to be the first reference to what we now refer to as "*particle size-selective*" exposure assessment for aerosols. By the early 1950s, there was sufficient information, experimental and theoretical, to enable consideration of the sizes of inhaled particles which were capable of penetrating to the alveolar region of the lung (as reviewed by Davies, 1952). In particular, this highlighted the significance of what is now known as "*particle aerodynamic diameter*" (d_{ae}) as distinct from the geometrical measure of particle size as obtained during the microscope counting that had previously been used.

Driven primarily by concerns about the high incidence of pneumoconiosis (a dust-related lung disease specifically associated with the alveolar region of the lung) among coal miners, this led in turn to the first definition of the convention that has been referred to since as the "respirable fraction." In 1952, the British Medical Research Council (BMRC) defined respirable aerosol in terms of the probability (on average, for a population of typical human subjects and for ranges of typical breathing and physiological parameters), as a function of d_{ae}, that an inhaled particle may penetrate to the alveolar region of the lung (BMRC, 1952; *see also* Hamilton and Walton, 1961). The BMRC definition is described by a curve where the penetration probability falls from unity at d_{ae} = 0 µm, passing through 50% at 5 µm and reaching zero at 7.1 µm. Although this curve is quite closely representative of what happens in humans, it is defined exactly by the physical penetration of particles through a laminar-flow horizontal elutriator of appropriate dimensions and flowrate. In such an elutriator, particles are separated by gravitational settling. Members of the BMRC Committee took the view that a definite method of separation resting on a firm physical basis should be specified. It should be similar to — but not necessarily in exact agreement with — the lung deposition data, bearing in mind that present information was imprecise and that there was considerable variation between individuals.

Essentially the same recommendations were later made during the 1959 Pneumoconiosis Conference in Johannesburg, and the definition later became well-known in some quarters as the "*Johannesburg curve*" (Orenstein, 1960). Later still, other curves were proposed for the respirable fraction, most notably the 1968 curve of the American Conference of Governmental Industrial Hygienists (ACGIH®, 1968), where the 50% penetration probability was placed at d_{ae} = 3.5 µm.

In the late 1970s, the International Standards Organization (ISO) formed an *ad hoc* working group to discuss further elaboration of the particle size-selective exposure assessment concept. Informal meetings were held in Gainesville, Florida and London, United Kingdom. From those meetings came a consensus that the framework for particle size-selective sampling should be expanded to cover all inhaled particles which might be related to all aerosol-related health effects. In particular, the idea emerged that three particle fractions should be defined: "*inhalable*"* ((relating to all particles that are capable of entering through the nose and/or mouth during breathing), "*thoracic*" (relating to all inhaled particles that may penetrate to the lung below the larynx), and "*respirable*" (defined as in the earlier BMRC and ACGIH conventions to relate to all inhaled particles that may penetrate to the alveolar region of the lung). Based on the available data from experiments with mannequins in wind tunnels and human subjects in the laboratory, ISO proposed the first set of quantitative criteria for each of these fractions, taking the form — as it had done for the earlier conventions — of the probability of penetration as a function of d_{ae} (ISO, 1983 and 1992). The scope of the ISO recommendations included not only workplace aerosol exposures but also exposures of the general population in the ambient air.

In the early 1980s, ACGIH formed its Air Sampling Procedures Committee to examine the same matter, focusing primarily on aerosol exposures in workplaces, and aimed at identifying a rationale basis for *Threshold Limit Values* (TLVs®, which are the OELs recommended by ACGIH). The 1985 report of this Committee (Phalen, 1985) is the forerunner of this book. Although it identified the same three aerosol fractions as had been described by ISO, there were some quantitative and conceptual differences, arising from the application of new data not available to the ISO working group as well as some differences in the way the data were interpreted and applied. By the late 1980s, however, there was strong momentum towards achieving international harmonization of the concepts of particle size-selective aerosol sampling. A set of criteria was agreed upon by some of the world's leading standards setting bodies, notably ISO (1992), ACGIH (1992 onwards, as reflected in ACGIH, 1998) and the Comité Européen de Normalisation (CEN, 1992).

---

* A note is required about terminology. In its original deliberations, members of the ISO ad hoc working group adopted the word "inhalable." But this was later changed to "inspirable" in order to avoid possible confusion with other usages of the term "inhalable," most notably by the U.S. Environmental Protection Agency (EPA) to relate to what we now refer to as the thoracic fraction. Later, however, EPA adopted the term "PM_{10}", which is retained to this day. This then cleared the way for a change back to the more appropriate term "inhalable" as it is used throughout this book.

Over the years, therefore, there appears to have evolved a harmonized set of particle size-selective criteria for aerosol exposure assessment which has a firm scientific basis and upon which most of the interested scientific parties are able to agree (Soderholm, 1989). In addition, however, new issues are now starting to emerge which suggest that further expansion of the set of criteria may be required. In the meantime, however, stimulated by — and in parallel with — the development of the criteria, there has been a growth in the activity to develop and make commercially available new sampling instruments with performance characteristics matching those criteria.

With the preceding in mind, it is now timely to re-evaluate and consolidate the scientific base of knowledge from which these now widely-agreed particle size-selective criteria have emerged. Such consolidation will clear the path towards the deployment of new generations of sampling methods that provide aerosol exposure measurements more relevant to health and the development of improved and more-relevant standards. This book aims to provide that scientific base.

The book is presented in two parts. The first covers the background and status of what now is generally agreed and adopted. Chapter 1 sets out the scientific rationale for particle size-selective aerosol sampling in both working and living environments and Chapter 2 provides the essential physiological background. Chapters 3 and 4 describe the research leading up to the development of formal definitions for the inhalable, thoracic and respirable conventions aimed primarily at the working environment; Chapter 5 describes the development of corresponding particle size-selective conventions for particles in the ambient atmospheric environment. Chapters 6 to 8 describe the history and current status of practical sampling instrumentation for the measurement of the various particle size fractions. Chapter 9 draws together the contents of the preceding chapters and summarizes what has now been officially adopted by ACGIH as the basis of its TLVs for substances occurring as aerosols. Finally in this part of the book, Chapter 10 reviews the general framework for the development of the ACGIH TLVs and discusses how the new particle size-selective sampling criteria may be applied in that process.

The second part of the book deals with emerging issues where new knowledge is pointing the way towards the development of new or extended particle size-selective criteria. Chapter 11 draws the distinction between particles that penetrate into the lung and those which are actually deposited, and how this might be important in the description of inhaled dose in certain practical situations. Chapter 12 discusses the topical question of particles that are so small their behavior in terms of how they react with biological systems may be very different from that for larger particles of the same material. Finally, Chapter 13 addresses the important practical question of how standards should be set by which to define and determine the acceptability of aerosol sampling instruments in relation to the new particle size-selective criteria. This is an area where, to date at least, most of the running has been made in Europe.

The editor and the authors, representing the ACGIH Air Sampling Procedures Committee, wish to thank everyone who has contributed to this discussion over many years. In particular, thanks are due to the ACGIH Board of Directors and the hardworking staff who have continued to support this activity since the early 1980s and encouraged the progress which has been achieved towards new TLVs based on particle size-selective sampling criteria. Thanks are also due to the members of the Chemical Substances TLV Committee who have encouraged this work for many years, recognizing at an early stage its importance to the development of more rational aerosol standards.

James H. Vincent
Editor and Chair, ACGIH Air Sampling
Procedures Committee

REFERENCES

American Conference of Governmental Industrial Hygienists (ACGIH) (1968), Threshold limit values of airborne contaminants, ACGIH, Cincinnati, OH.

American Conference of Governmental Industrial Hygienists (ACGIH) (1998), Threshold limit values for chemical substances and physical agents, and biological exposure indices, ACGIH, Cincinnati, OH.

British Medical Research Council (BMRC) (1952), Recommendations of the BMRC panels relating to selective sampling, From the Minutes of a joint meeting of Panels 1, 2 and 3 held on March 4, 1952 (see also Hamilton and Walton, 1961).

Comité Européen de Normalisation (CEN) (1992), Workplace atmospheres: size fraction definitions for measurement of airborne particles in the workplace, CEN Standard EN 481.

Davies, C.N. (1952), Dust sampling and lung disease, *Br. J. Ind. Med.*, 9, pp. 120-126.

Hamilton, R.J. and Walton, W.H. (1961). The selective sampling of respirable dust. In: *Inhaled Particles and Vapours* (C.N. Davies, Ed.), Pergamon Press, pp. 465-475.

International Standards Organization (ISO) (1983 and 1992), Air quality — particle size fraction definitions for health-related sampling, Technical Report ISO/TR/7708-1983 (E), ISO, Geneva, originally 1983, revised 1992.

McCrea, J. (1913), The ash of silicotic lungs, Publication of South African Institute of Medical Research, Johannesburg.

Orenstein, A.J. (Ed.) (1960), Recommendations adopted by the Pneumoconiosis Conference, In: *Proceedings of the Johannesburg Pneumoconiosis Conference* (A.J. Orenstein, Ed.), Churchill, London, pp. 619-621.

Phalen, R.F. (Ed.) (1985), Particle size-selective sampling in the workplace, Report of the Air Sampling Procedures Committee, American Conference of Governmental Industrial Hygienists (ACGIH), Cincinnati, OH.

Soderholm, S.C. (1989), Proposed international conventions for particle size-selective sampling, *Ann. Occup. Hyg.*, 33, pp. 301-321.

Walton, W.H. and Vincent, J.H. (1998), Aerosol instrumentation in occupational hygiene: an historical perspective. *Aerosol Sci. Tech.*, 28, pp. 417-438.

Postscript

As this book goes to press, it is clear that aerosol science as it relates to occupational and environmental health is a continually evolving field. So this book does not represent the last word on the subject, but rather another milestone on the road to the application of good science in improved standards. No doubt the ideas and resultant criteria that appear in the first part of the book will continue to evolve and improve, and the new emerging issues that appear in the second part will grow and be developed further. It is also in the nature of science that new such issues will continue to surface, many of which will be at least as important as the ones that came before. The aerosol science and the environmental and occupational health science communities need, therefore, to remain vigilant and ready to rise to the new challenges that lie ahead. The ACGIH Air Sampling Procedures Committee will continue to be part of that effort.

James H. Vincent
Editor and Chair, ACGIH Air Sampling
Procedures Committee

PART I

BACKGROUND TO AND DEVELOPMENT OF ADOPTED PARTICLE SIZE-SELECTIVE AEROSOL SAMPLING CONVENTIONS

Chapter 1

RATIONALE FOR PARTICLE SIZE-SELECTIVE AEROSOL SAMPLING

Morton Lippmann

Nelson Institute of Environmental Medicine,
New York University School of Medicine

1.1 Introduction

The major goal of particle size-selective sampling in working and living environments is to provide the most appropriate index of particle inhalation hazards by giving recognition to the fact that the particle size characteristics can greatly modify their regional deposition.

Historically, aerosol sampling in workplaces for the purpose of assessing the exposures of workers goes back to the turn of the twentieth century (Walton and Vincent, 1998). In the early 1900s, relatively short-period gravimetric samples (for nominally 'total dust') were taken, usually at breathing height near representative workers during active operations, primarily to locate sources and to test the effectiveness of dust control measures (and requiring the involvement of an attendant operator). From about 1920, following realization that the mass of the non-respirable large particles in the 'total dust' sampled over-estimated the hazard associated with some types of lung disease (e.g., coalworkers' pneumoconiosis), short-period or 'snap' sampling for microscopical counts of particle numbers of the more relevant fine particles (usually those with geometrical size less than 5 μm) was widely carried out. From about 1950, increasing emphasis was given to full-shift time-weighted average (TWA) particle-number sampling (usually by multiple short-period samples), either near representative workers as a measure of health-related exposure or at strategic sampling positions to monitor dust control (still requiring attendance of an operator). Then in the 1960s, development took place of full-shift, TWA, positional (area or static) samplers that employed aerodynamic selection of fine, so-called *'respirable'* particles, were able to operate unattended and provided samples which could be assessed in terms of mass or chemical (including mineralogical) composition. By this time, it had been realized that respirable particulate mass concentration generally provided a better index of most pneumoconiosis risk than did particle number concentration. This concept was applicable not only to the fine respirable fraction, but also to the coarser aerosol which remains more relevant to some types of aerosol-related ill health associated with effects in the larger conductive airways. From the 1970s onwards, increasing emphasis has been given to TWA personal sampling for particulate mass because this is the approach considered by most professional industrial hygienists as providing the best estimate of actual worker exposure most relevant to disease risks.

As the preceding short history indicates, for workers in the occupational setting, particle size-specific mass began to supplement or replace total sampled mass for certain types of aerosol when it became accepted that the use of total mass concentration ignores the fact that toxicants captured at various sites in the upper respiratory tract or tracheobronchial tree may at times control the extent of the hazard.

As far as the aerosol exposure of the general population is concerned, however, the first official recognitions that total sampled mass may not be the best index of biologically-relevant concentration in community air came much later — by the U.S. Environmental Protection Agency (EPA) in 1979 (Miller *et al.*, 1979) and by the International Standards Organization (ISO) in 1983. The 1983 ISO criteria were the first to address multiple particle size-selective criteria (in the forms of inhalable, thoracic, and respirable fractions) and included separate respirable dust criteria for occupational and community exposures. ISO divided particles into a number of fractions which included the respirable dust fraction penetrating to the gas exchange region, the thoracic fraction (which includes penetration to the tracheobronchial region as well as the gas exchange region), and the head airways region. Collectively, these fractions, involving not only the deposited aerosol but also that which is exhaled, constitute the inhalable (a.k.a., inspirable) fraction.

When the composition of the particles is known and relatively constant, then aerodynamic separation of particles into the inhalable, thoracic and/or respirable particulate fractions and their subsequent gravimetric analysis may be sufficient for most exposure assessment purposes. But when the composition of the aerosol varies with particle size, then exposure assessment for the component of interest or concern needs to be specific for specific constituents. In some cases, however, separate analyses of several components within one size fraction may not be feasible. One such example is in underground coal mines that employ diesel powered equipment for personnel and/or ore transport. Both the diesel engine exhaust and the coal mine dust are heavily dominated by fixed carbon, and it may be important to know the mass concentration attributable to each source. In this case, the particles emitted by diesel engines are almost all smaller than 0.8 μm, while the dust particles arising from coal cutting and hauling are almost all larger. Thus, the mass concentration of fixed carbon below 0.8 μm can be attributed to diesel exhaust, while the concentration of fixed carbon between 0.8 and up to about 10 μm represents the coal dust fraction of the respirable coal mine dust (Rubow *et al.*, 1990).

With the exception of special cases, such as diesel exhaust in coal mines, the aerodynamic particle size fractions associated with the specified deposition regions are considered to be valid for occupational exposures. On this basis, it was a major initial charge to the ACGIH Air Sampling Procedures Committee (in 1982) to define particle size-selective aerosol sampling equipment specifications and procedures required to evaluate potentially hazardous aerosol-related situations in the workplace. A framework for establishing threshold limit values (TLVs) with size discrimination for individual compounds was outlined in the 1985 report of the Committee (Phalen, 1985) in terms of the physiological or biochemical responses or the pulmonary diseases associated with each. With the application of particle size-selective sampling criteria, future proposed and revised TLVs for substances occurring as aerosols would take into account not only the inherent toxicity of the particles, but also their particle size distributions, their patterns of deposition within the respiratory tract, and the particle size-related rate of dissolution and translocation to target tissues. Since 1985, the specific particle size-selective specifications have been modified in the interests of achieving international harmonization and in the light of new scientific knowledge, as discussed later in this chapter. But the broad concepts have remained essentially the same.

For ambient atmospheric aerosols, the EPA has adapted two different rationales for its particle size-selective sampling criteria. In its 1987 revision of the National Ambient Air Quality Standard (NAAQS) for ambient atmospheric particulate matter (PM), EPA switched its basis from total suspended particulate matter (TSP) to thoracic particulate matter, as measured in terms of PM_{10}, on the grounds that the primary health effects associated with PM exposure were limited to those attributable to particles depositing in the thorax (tracheobronchial and gas exchange airways). Thus the PM_{10} sampling criterion represented a conservative approximation of the ambient particles that could penetrate through the head airways to the thorax (Miller *et al.*, 1979). In the more recent NAAQS review process for PM, it became evident that the adverse health effects associated with exposure to PM of outdoor origin were more closely associated with the fine particles in the accumulation mode, as measured by $PM_{2.5}$, than with the larger, coarse-mode particles within PM_{10} (EPA 1996 a and b). On this basis, the EPA Administrator promulgated new daily maximum and annual average $PM_{2.5}$ NAAQS in July 1997 to supplement the existing PM_{10} NAAQS (EPA, 1997). Since the distinction between $PM_{2.5}$ and the coarse-mode particles within PM_{10} is based on the assumption of different sources and chemical compositions rather than on regional lung deposition, EPA elected to require as sharp a 'cut' at $d_{ae} = 2.5$ µm as was technically feasible to minimize the intrusion of coarse mode particles into the fine particle sample and vice-versa (Wolff, 1996).

The effect of utilizing one or other of these criteria on the mass concentration measured using a particle size-selective inlet on a particle sampler is illustrated in terms of particle size versus aerosol mass distribution in Figure 1.1. This distribution is for a hypothetical mixed aerosol containing both mechanically-generated dust particles that are generally larger than about 1 µm, and finer particles resulting from gas-phase chemical conversions of precursor vapors and/or condensation of fume particles with particle aggregation and hygroscopic growth leading to particle or droplet diameters up to, but rarely exceeding, 2 to 3 µm.

1.2 Review of Standards and Criteria for Particle Size-Selective Dust Samplers

Particle size-selective occupational exposure limits have already been established to address the problems associated with pneumoconiosis, and the development of standards for respirable dust are reviewed in this subsection. However, the limitations have not all been

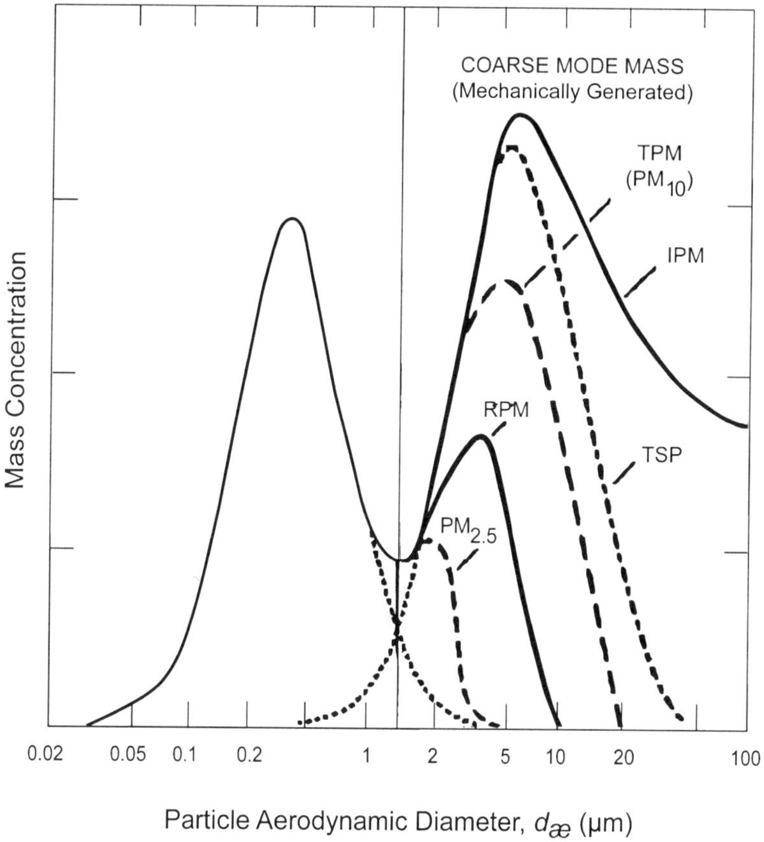

FIGURE 1.1. Effect of size-selective inlet characteristics on the aerosol mass collected by a downstream filter. IPM = inhalable particulate matter; TSP = total suspended particulate; TPM = thoracic particulate matter (aka PM_{10}); RPM = respirable particulate matter; and $PM_{2.5}$ = fine particulate matter in ambient air.

addressed because the information used to develop the standards was, in most cases, quite limited.

British Medical Research Council (BMRC)

In 1952 the British Medical Research Council (BMRC) adopted a definition of "*respirable dust*" applicable to pneumoconiosis producing dusts, referring it to inhaled particles that reach the alveolar region. BMRC selected the horizontal elutriator as a practical aerodynamic size selector and quantitatively defined respirable dust in terms of the penetration of an ideal horizontal elutriator in which the elutriator 'cut' characteristics provided the best match to experimental human respiratory tract data that were available at that time. The same standard was adopted by the Johannesburgh International Conference on Pneumoconiosis in 1959 (Orenstein, 1960). In order to implement these recommendations, it was specified that:

1. The estimation of airborne dust in its relation to pneumoconiosis, compositional analysis, or assessment of concentration by a bulk measurement such as that of mass or surface area, would involve only the "respirable" fraction of the cloud.
2. The "respirable" samples should be separated from the cloud while the particles are airborne and in their original state of dispersion.
3. The "respirable fraction" was defined in terms of the penetration of particles through a horizontal elutriator based on the deterministic theory of Walton (1954). Specifically, it was defined as a penetration curve having the form

$$P(d_{ae}) = 1 - \frac{v_s}{2v_{s,5}}$$

in which v_s is the settling velocity of a particle of aerodynamic diameter d_{ae} and $v_{s,5}$ is the settling velocity of a particle with d_{ae} = 5 µm. Thus it is seen that P passes through 50% at d_{ae} = 5 µm and reaches zero at d_{ae} = 7.1 µm when $v_s = 2v_{s,5}$.

U.S. Atomic Energy Commission (AEC)

A second standard for *"respirable dust"* was established in January 1961 at a meeting sponsored by the U.S. Atomic Energy Commission (AEC), Office of Health and Safety (Hatch and Gross, 1964). It referred to respirable aerosol as those portions of the inhaled dust which penetrate to the non-ciliated portions of the gas exchange region. This application of the respirable dust concept and concomitant selective sampling was intended only for "insoluble" particles which exhibit prolonged retention in the lung. It was not intended to include dusts which have an appreciable solubility in body fluids and those which are primarily chemical toxicants. Within these restrictions, "respirable dust" was defined in terms of a penetration curve with a 50% 'cut' at d_{ae} = 3.5 µm, also passing through 100% at d_{ae} = 2 µm, 75% at 2.5 µm, 50% at 5 µm and 0% at 10 µm.

AMERICAN CONFERENCE OF GOVERNMENTAL INDUSTRIAL HYGIENISTS (ACGIH)

CRYSTALLINE SILICA

ACGIH, at its annual meeting in St Louis, Missouri, in May 1968, announced in their "Notice of Intended Changes" alternate mass concentration TLVs for quartz, cristobalite and tridymite (three forms of crystalline free silica) to supplement the TLVs based on particle number count concentrations. For quartz, the alternative mass values proposed were (ACGIH, 1968):

1. For respirable dust:

$$\text{TLV} = \{10 \ / \ (\% \text{ respirable quartz} + 2)\} \text{ mg/m}^3$$

where both concentrations and % quartz for the application of this limit are to be determined from the fraction that penetrates a size-selective sampler with the following characteristics:

d_{ae} (μm)	2.0	2.5	3.5	5.0	10
% Penetration	90	75	50	25	0

2. For 'total' dust (i.e., both respirable and non-respirable):

$$\text{TLV} = \{30 \ / \ (\% \text{ quartz} + 2)\} \text{ mg/m}^3$$

3. For cristobalite and tridymite:

TLV = One-half the value calculated from the count or mass formulae for quartz.

The size-selector characteristic specified by ACGIH was almost identical to that of the AEC, differing only at $d_{ae} = 2$ μm where it allows for 90% passing the first stage collector instead of 100%. The difference appears to be a recognition of characteristics of real particle separators. For practical purposes, however, the two standards may be considered equivalent.

The proposed mass concentration limits were obtained by a comparison of simultaneous impinger and particle size-selected samples collected in Vermont granite sheds (Sutton and Reno, 1968). Since the original impinger sampling and microscopic particle counting standards were based on epidemiological investigations that had been performed three to four decades earlier in some of the same granite cutting sheds, it was possible to make a valid comparison of "respirable" mass and particle count exposure data.

COAL DUST

The following values were suggested for coal dust:

1. For respirable dust:

   TLV = 2 mg/m^3 (respirable particulate mass concentration, or RPMC, with less than 5% quartz).

   or

   TLV = {10 / (% quartz + 2)} if the respirable dust fraction contained greater than 5% quartz.

The coal dust standard was not well defined for all of the potential coal dust-related diseases. It was developed to address specifically the problem of pneumoconiosis, but it did not deal with the industrial exposures leading, for example, to immunological effects and the onset of bronchitis.

INERT OR NUISANCE DUST

The following values were suggested for inert or nuisance dusts:

1. For 'total' dust:

   TLV = 30 million particles per cubic foot (mppcf) or 10 mg/m^3 of dust containing less than 1% quartz

2. For respirable dust:

   TLV = 5 mg/m^3

The Federal Coal Mine Health and Safety Act of 1969 specified that

> *References to concentrations of respirable dust in this title means the average concentration of respirable dust if measured with an MRE instrument or such equivalent concentrations if measured with another device approved by the Secretary (of Interior) and the Secretary of Health, Education and Welfare. As used in this title, the term "MRE" instrument means the gravimetric dust sampler with four channel horizontal elutriator developed by the Mining Research Establishment of the National Coal Board, London, England.*

While the 1969 Act specified a British instrument, which follows the BMRC sampling criterion (the so-called "MRE"), the Federal Mine Safety and Health Act of 1977, which superseded it, does not. The National Research Council Committee on Measurement and Control of Respirable Dust in Mines (NRC, 1980) noted that it may be more appropriate to use the definition of respirable dust adopted by ACGIH, since the available human deposition data suggest that the ACGIH curve is a better representation of respirable dust than is the BMRC curve.

Under the Occupational Safety and Health Act of 1970, the Occupational Safety and Health Administration (OSHA) adopted 22 Maximum Acceptable Concentrations (MACs) of the American National Standards Institute (ANSI) and approximately 280 of the ACGIH 1968 TLVs including the silica TLVs which specify either dust counts or respirable mass concentrations. Since its establishment, OSHA has adopted only a few comprehensive health standards, none of which address the issue of respirable dust. The Mine Safety and Health Administration (MSHA) of the Department of Labor operates under different enabling legislation and uses the 1973 TLVs which, for silica, are the same as the 1968 values.

In 1982, the ACGIH Board appointed an ad hoc Committee on Air Sampling Procedures to prepare general recommendations for size-selective sampling appropriate to particle size-selective TLVs (PSS-TLVs) for particulate materials. This Committee had as its primary charge

> *. . . to recommend size-selective aerosol sampling procedures which will permit reliable collection of aerosol fractions which can be expected to be available for deposition in the various major subregions of the human respiratory tract, e.g., the head, tracheobronchial region, and the alveolar (pulmonary) region.*

It was anticipated from the outset that the work of this Committee would lead to an approach for establishing PSS-TLVs for many airborne agents. The Committee reviewed the relevant literature and the recommendations of other groups on particle size-selective aerosol sampling; its report and recommendations were presented to the Board of Directors and the membership of ACGIH at the 1984 Annual Meeting. The initial report of the Committee and its background documentation were published in the *1984 Annals of ACGIH* (ACGIH, 1984) and appeared in 1985 as a separate document entitled *Particle Size-Selective Sampling in the Workplace* (Phalen, 1985) which was the forerunner to this book.

This and the following paragraphs summarize the recommendations which appeared in the 1985 report of the Committee. This report was a background document summarizing the available data on: a) airway anatomy and physiology which influence the deposition and retention of inhaled particles; b) penetration of inhaled particles into the major functional regions of the respiratory tract; c) the particle size collection characteristics of available size-selective aerosol samplers; and d) evaluation of the performance of samplers. It also reviewed the basis for its particular recommendations on size-selective sampling criteria and how and why they differed from the recommendations of others.

The major functional regions of the human respiratory tract were given different names and/or abbreviations than those used by others but were anatomically equivalent, as indicated in Table 1.1. The designations chosen were, in the Committee's view, more anatomically correct and unambiguous. Deposition within the head airways region (HAR) was known to be associated with an increased incidence of nasal cancer in wood and leather workers and in ulceration of the nasal septum in chrome refinery workers. Within the tracheobronchial region (TBR), deposited particles can contribute to the pathogenesis of bronchitis and bronchial cancer. Particles depositing within the gas exchange region (GER) can cause emphysema and fibrosis. Particles that can enter the systemic circulation from a specific region of the respiratory tract should be assigned a PSS-TLV appropriate to that region. But more generally, the hazards from inhaled materials that exert their toxic effects on critical sites outside the respiratory tract, after dissolution into circulating fluids, depend upon total respiratory tract deposition rather than deposition with one region.

The Committee considered several options for size-selective sampling of fractions of the aerosol that represent hazards for specific health endpoints. The major options were: a) samplers that would mimic deposition in the specific regions of interest; and b) samplers that would collect those particles which would penetrate to, but not necessarily deposit in, the specific region of interest. The Committee opted for the latter approach as the one requiring simpler and less expensive samplers and therefore being more practical. This approach had already proven to be effective in sampling for respirable dust. There it had been recognized that measured respirable dust concentrations may be as much as five to ten times greater than the fraction actually depositing in the lungs (depending on particle size distribution), since up to 75% of particles in the 0.1 to 1.0 μm diameter range are exhaled. However, because the fraction deposited in the

TABLE 1.1. Respiratory tract regions as defined in particle deposition models

ACGIH Region	Anatomic Structure Included	ISO Region	1966 ICRP Task Group Region	1994 ICRP Task Group Region	1997 NCRP Task Group Region
1. Head Airways (HAR)	Nose Mouth Nasopharynx Oropharynx Laryngopharynx	Extrathoracic (E)	Nasopharynx (NP)	Anterior Nasal Passages (ET_1) All other Extrathoracic (ET_2)	Naso-oropharyngo-laryngeal (NOPL)
2. Tracheobronchial (TBR)	Trachea Bronchi Bronchioles (to terminal bronchioles)	Tracheobronchial (B)	Tracheobronchial (TB)	Trachea and Large Bronchi (BB) Bronchioles (bb)	Tracheobronchial (TB)
3. Gas Exchange (GER)	Respiratory bronchioles Alveolar ducts Alveolar sacs Alveoli	Alveolar (A)	Pulmonary (P)	Alveolar Interstitial (AI)	Pulmonary (P)

GER is a relatively constant one over the whole "respirable dust" size range, the measured respirable dust concentration was considered to be an adequate index of the hazard for mineral dusts. It was recognized that a sampler which would mimic GER deposition would be much more difficult to design and operate, and it would not necessarily give a materially better index of mineral dust inhalation hazard.

The aerosol that enters the HAR was referred to in the Committee's report as "*inspirable particulate mass (IPM)*" fraction (renamed "*inhalable*" in 1993). The aerosol that penetrates the HAR and enters the TBR was referred to as the "*thoracic particulate mass (TPM)*" fraction. Here, the Committee chose to define the thoracic fraction on the basis of data for HAR deposition during mouth breathing. The difference between the IPM and the TPM fractions approximates the deposition fraction in the HAR occurring during mouth breathing. Since nasal inhalation would almost always produce more HAR deposition than oral inhalation, actual HAR deposition during nasal breathing would be greater than that calculated. Similarly the TPM fraction as defined would overestimate the hazard to the TBR region for nose-breathing workers. The Committee's selection of a mouth-breathing model rather than a nose-breathing model was therefore made in order to be conservative. Occupational diseases of the lung airways are much more common than are diseases of the head airways. Also, heavy work in industry is believed to cause a significant fraction of workers to engage in mouth breathing during periods of maximal activity, which may coincide with maximal levels of airborne dust. The algebraic difference between TPM and respirable particulate mass (RPM) approximates tracheobronchial region deposition during oral breathing. By contrast, for nasal breathing individuals, the difference between TPM and RPM is a much poorer estimate of tracheobronchial deposition.

In general, mass concentrations tend to be dominated by the largest particle size-fraction collected. In consideration of all these factors, the Committee recommended that samplers which follow its IPM criterion be used for sampling those materials that are hazardous when deposited in the HAR or when systemic toxicity can follow from deposition anywhere in the respiratory tract. For those materials that represent a hazard when deposited on the conductive airways of the lungs, it was recommended to use a sampler that follows the TPM criterion. Finally, for those materials, such as silica, which are hazardous only after deposition in the GER, the Committee recommended using a sampler which follows its RPM sampling criterion.

The Committee's recommendations for the performance specifications of samplers that would mimic aerosol penetration into these regions were similar, but not identical, to the original recommendations of ISO (1983). The most notable differences were in the IPM criteria and the RPM criteria. In terms of the former, the Committee had the advantage over ISO of access to inhalability data for particles with d_{ae} larger than 40 μm that were not available to the original ISO Working Group. That ISO Group had made the reasonable, but inadequate, assumption that the data for $d_{ae} < 40$ μm could be extrapolated to zero deposition at $d_{ae} = 185$ μm. For RPM, the major difference was in not having the alternate criteria based on the BMRC recommendations.

The Committee's recommendations for sampling TPM were quite similar to the original recommendation of ISO and the EPA. The recommendations also contained sampler acceptance envelopes about the recommended curves.

Following the ACGIH Board's acceptance of the Committee's recommendations in 1984, activities to implement the recommendations proceeded in two ACGIH committees, the Chemical Substances TLV Committee and the Air Sampling Procedures Committee. The Chemical Substances TLV Committee addressed the use of the particle size-selective criteria for airborne particulate matter by listing the criteria as an issue under study in the *Theshold Limit Values and Biological Exposure Indices* (TLV/BEI) booklet for 1986–1987. In the following year, ACGIH adopted these criteria as a separate appendix in the annual TLV/BEI booklet. This appendix, reproduced below, incorporated the original terminology, nomenclature, sampling definitions and rationale recommended by the Air Sampling Procedures Committee, as follows:

> *For chemical substances present in inhaled air as suspensions of solid particles or droplets, the potential hazard depends on particle size as well as mass concentration because of: 1) effects of particle size on deposition site within the respiratory tract, and 2) the tendency for many occupational diseases to be associated with material deposited in particular regions of the respiratory tract.*
>
> *ACGIH has recommended particle size-selective TLVs for crystalline silica for many years in recognition of the well-established association between silicosis and respirable mass concentrations. It now has embarked on a re-examination of other chemical substances encountered in particulate form in occupational environments with the objective of defining: 1) the*

size-fraction most closely associated for each substance with the health effect of concern; and 2) the mass concentration within that size fraction which should represent the TLV.

The Particle Size-Selective TLVs (PSS-TLVs) will be expressed in three forms, e.g.,

a. *Inspirable Particulate Mass TLVs (IPM-TLVs) for those materials which are hazardous when deposited anywhere in the respiratory tract.*
b. *Thoracic Particulate Mass TLVs (TPM-TLVs) for those materials which are hazardous when deposited anywhere within the lung airways and the gas-exchange region.*
c. *Respirable Particulate Mass TLVs (RPM-TLVs) for those materials which are hazardous when deposited in the gas-exchange region.*

The three particulate mass fractions described above are defined in quantitative terms as follows:

a. *Inspirable Particulate Mass consists of those particles that are captured according to the following collection efficiency regardless of sampler orientation with respect to wind direction:*

$$\mathrm{IPM}(d_{ae}) = 0.5 \ \{1 + \exp(-0.06\ d_{ae})\} \pm 0.1; \quad \text{for } 0 < d_{ae} \leq 100\ \mu\text{m}$$

Collection characteristics for $d_{ae} > 100\ \mu m$ *are presently unknown. In this equation,* d_{ae} *is aerodynamic diameter in* μm.

b. *Thoracic Particulate Mass consists of those particles that penetrate a separator whose size collection efficiency is described by a cumulative lognormal function with a median* $d_{ae} = 10\ \mu m \pm 1.0\ \mu m$ *and with a geometric standard deviation* (σ_g) *of 1.5 (± 0.1).*
c. *Respirable Particulate Mass consists of those particles that penetrate a separator whose size collection efficiency is described by a cumulative log-normal function with a median* $d_{ae} = 3.5\ \mu m \pm 0.3\ \mu m$ *and with a* σ_g *of 1.5 (± 0.1). This incorporates and clarifies the previous ACGIH Respirable Dust Sampling Criteria.*

These definitions provide a range of acceptable performance for each type of size-selective sampler. Further information is available on the background and performance criteria for these particle size-selective sampling recommendations.

In 1989, Soderholm, then Chair of the Air Sampling Procedures Committee, with the endorsement of the full Committee, proposed modified particle size-selective sampling criteria for adoption by ACGIH, ISO, and the Comité Européen de Normalisation (CEN), with the objective of easing the path towards international harmonization (Soderholm, 1989). This initiative was well received by the interested parties and is being implemented by many of those concerned (ISO, 1992; CEN, 1992; ACGIH, 1994). Those proposals were adopted by ACGIH itself in 1993 and so now represent the latest ACGIH recommendations for workplace particle size-selective sampling. The now widely accepted definitions for the inhalable, thoracic, and respirable fractions are as summarized in Chapter 9. Penetration efficiencies for ranges of particle sizes in each of the fractions are shown in Table 1.2. Full details of the scientific bases supporting these definitions are given in Chapters 3 to 5.

An issue initially raised in Appendix F of the *1986-87 TLV/BEI* booklet concerned the changing of all TLVs that were explicitly defined in terms of "*total dust*" to "*inspirable particulate mass,*" without changing the numerical values. It should be noted, however, that the only explicit references to 'total dust' in the current TLVs are for mineral dusts. At its 1993 Annual Meeting, ACGIH endorsed the need to examine each TLV for airborne particles for conversion to PSS-TLVs. Since then, a number of substances occurring as aerosols have been specified in new or proposed new TLVs in terms of the inhalable fraction.

SIZE-SELECTIVE CRITERIA FOR SPECIFIC OCCUPATIONAL HAZARDS

Separate criteria have been adopted for cotton dust (NIOSH, 1975) and asbestos (Leidel *et al.*, 1979; ACGIH/AIHA, 1975; Health Effects Institute [HEI], 1991). However, these are different from the ones in the previous section. So a brief review of the rationale and practice for each of these materials is given below.

COTTON DUST SAMPLING

Since byssinosis, or so-called "brown lung," is characterized by an allergic response producing airway constriction, it was recognized that

TABLE 1.2. Inhalable, thoracic, and respirable dust criteria of ACGIH-ISO-CEN.

Inhalable		Thoracic		Respirable	
Particle Aerodynamic Diam. (μm)	Inhalable Particulate (%)	Particle Aerodynamic Diam. (μm)	Thoracic Particulate (%)	Particle Aerodynamic Diam. (μm)	Respirable Particulate (%)
0	100	0	100	0	100
1	97	2	94	1	97
2	94	4	89	2	91
5	87	6	80.5	3	74
10	77	8	67	4	50
20	65	10	50	5	30
30	58	12	35	6	17
40	54.5	14	23	7	9
50	52.5	16	15	8	5
100	50	18	9.5	10	1
		20	6		
		25	2		

particles depositing in the tracheobronchial airways should not be excluded. Thus, conventional "respirable" dust criteria were judged to be inappropriate. On the other hand, the mass of the dust in cotton ginning and textile operations tends to be dominated by very large cotton fibers which are too large to be inhalable. These considerations led to the recommendation of a standard that requires the use of a preseparator which consists of a vertical elutriator with a penetration having a nominal 50% 'cut' at d_{ae} = 15 μm (NIOSH, 1975). The second-stage filter is analyzed for the mass concentration of the particles judged most likely to be related to the health effects.

Asbestos Sampling

For asbestos and other durable mineral fibers, size-selection is generally applied after air sampling by restricting the microscopic analyses to fibers with geometric diameters < 3 μm. There is no aerodynamic size selectivity in the air sampling procedures specified in the NIOSH or ACGIH/AIHA sampling recommendations for occupational exposure, or by EPA for general population exposures in public and commercial buildings, although the specified inlet configurations to the filter holders will, of course, impose some. Fang and Lippmann (1995) have proposed an inlet design with a sharp 'cut' at

d_{ae} = 10 µm that would prevent fibers with physical diameters ≤ 3 µm from penetrating to the filter, thereby reducing extraneous debris from the microscopic analyses. In the analyses by phase contrast optical microscopy, there is an effective lower limit for fiber diameter imposed by the resolving power of the optical system. There are also other limits specified by the methods, whereby particles with aspect ratio (length to diameter) of less than 3, or a length of less than 5 µm, are not counted. These exclusions can be rationalized on the basis of toxicological and epidemiological studies which show that the toxic effects are primarily associated with long thin fibers. Asbestosis, a fibrotic disease, and mesothelioma, a cancer of the pleural or peritoneal surfaces, are presumably related to long fibers depositing in the gas exchange regions. Bronchial cancer may be related to the long fibers depositing on bronchial airways (Lippmann, 1988).

1.3 Regional Deposition and Clearance Dynamics

It is useful at this point to summarize briefly the biological and physiological mechanisms by which the human body defends itself against the challenge of inhaled particles. A fuller description is given in Chapter 2.

Definitions

Deposition: Refers specifically to the collection of inhaled airborne particles by the respiratory tract and to the initial regional patterns of these deposited particles.

Penetration: Except where otherwise defined, this refers to the aerodynamic passage of particles into the respiratory tract.

Clearance: Refers to the subsequent translocation, transformation and removal of deposited particles from the respiratory tract.

Retention: Refers to the temporal distribution of uncleared material.

For the purpose of estimating toxic dose from inhaled particles, the respiratory tract can be divided into a number of functional regions, which differ grossly from one another in deposition efficiencies, retention times at the deposition sites, and to some extent pathologic response. These are:

1. *Head airways region*
    a. Anterior unciliated nares (for nose breathing).
    b. Ciliated nasal passages, olfactory epithelium and nasal pharynx (for nose breathing).

c.1. Oral cavity, pharynx and larnyx (for mouth breathing).

c.2. Nasopharynx, pharynx and larynx (for nose breathing).

2. *Tracheobronchial region* (for both nose and mouth breathing).

3. *Gas exchange region* (for both nose and mouth breathing).

The fractional deposition in each of these regions is dependent on particle aerodynamic size and the subject's airway dimensions and breathing characteristics (flowrate, breathing frequency, tidal volume, etc).

Air sampling data collected using the procedures recommended by this Committee could provide data on the particle penetration to be expected into each functional region. However, at a minimum, the sampling should be selective for regions 1, 2, and 3, inclusive.

HEAD AIRWAYS REGION

NASAL PASSAGES

Air enters through the nares or nostrils, passes through a web of nasal hairs, and flows posteriorly toward the nasopharynx while passing through a series of narrow passages winding around and through shelf-like projections called turbinates. The air is warmed and moistened in its passage and partially depleted of particles. Some particles are removed by electrostatic forces and impaction onto the nasal hairs and at bends in the air path, and others by sedimentation and diffusion. Except for the anterior nares and olfactory region, the surfaces are covered by a mucous membrane composed of ciliated and goblet cells. The mucus produced by the goblet cells is propelled toward the pharynx by the beating of the cilia, carrying deposited particles along with it. Particles deposited on the anterior unciliated portion of the nares and at least some of the particles deposited on the nasal hairs usually are not carried posteriorly to be swallowed; rather, they are removed mechanically by nose wiping, blowing, sneezing, etc.

ORAL PASSAGES, PHARYNX AND LARYNX

During mouth breathing, some inhaled particles are deposited, primarily by impaction, in the oral cavity and at the back of the throat. These particles are rapidly eliminated to the esophagus by swallowing.

TRACHEOBRONCHIAL TREE

These airways taken collectively have the appearance of an inverted tree, with the trachea analogous to the trunk and the subdividing bronchi to the limbs. The branching pattern is nominally asymmetric in a regular pattern, as described by Horsfield *et al.* (1975). However, for purposes of discussion, it will be clearer if we adopt Weibel's simplified anatomic mode in which there are 16 generations of bifurcating ciliated airways (Weibel, 1963). The diameter decreases from generation to generation, but because of the increasing number of tubes, the total cross section for flow increases and the air velocity decreases toward the ends of the tree. In the larger airways, particles too large to follow the bends in the air path are deposited by impaction. At the low velocities in the smaller airways, particles deposit by sedimentation and, if small enough, by diffusion.

Ciliated and mucus secreting cells are found at all levels of the tracheobronchial tree. Within hours, inert non-soluble particles deposited in this region are thus carried towards the larynx on the moving mucus sheath which is propelled proximally by the beating of the cilia. Beyond the larynx, the particles enter the esophagus and pass through the gastrointestinal tract.

Persistent defects in clearance of particles from the bronchial tree, such as that caused by chronic exposure to cigarette smoke (Albert *et al.*, 1969) or sulfuric acid (Lippmann *et al.*, 1987), would also lead to increased residence times for particles containing toxic and carcinogenic chemicals. This increases the dose to the underlying tissues from those chemicals and results in increased systemic uptake. Consequently, defective clearance may contribute to a variety of disease conditions. In addition, some small particles may translocate through the mucus layer and so have prolonged retention (e.g., Gehr *et al.*, 1990).

GAS EXCHANGE (OR ALVEOLAR) REGION

The region beyond the terminal bronchioles, which includes the alveoli and associated ducts and bronchioles, is the region in which the gas exchange takes place. The epithelium is nonciliated and, therefore, insoluble particles deposited in this region by sedimentation and diffusion are removed at very slow rates, with clearance half-times on the order of one month or more. The mechanisms for particle clearance from this region are only partly understood and their relative importance is still a matter of some debate. Some particles are engulfed by phagocytic cells which are transported onto the ciliary "escalator" of the bronchial tree in an undefined manner.

Others move through the alveolar wall and enter the lymphatic system. Still others dissolve slowly *in situ*. Even supposedly "insoluble" particles have some finite dissolution rate, which is greatly enhanced for smaller particles by their large surface to volume ratio. Morrow *et al.* (1964) demonstrated that the clearance half-times of many "insoluble" dusts in the lung are proportional to their solubilities in simulated lung fluids. It has also been demonstrated that the gas exchange region clearance mechanisms may function differently for different dusts and concommitantly-inhaled gaseous contaminants. For very high levels of exposure, the lung clearance capacity can be exceeded, leading to the phenomenon know as "*dust overload.*" Under these conditions, an enhanced retention and an increased incidence of lung cancer in laboratory animals had been observed (Bolton *et al.*, 1981; Morrow, 1988). However, the significance of overload to humans is not yet understood.

In any case, it is likely that prolonged retention of particles in the gas exchange region of the lungs is non-beneficial. Prolonged retention of the inhaled particles in the gas exchange region increases the doses of those particles to the underlying tissues as well as the potential for systemic uptake. If the particles are fibrogenic, this prolonged retention could contribute to the development of pneumoconiosis and emphysema. Cigarette smoke contains a variety of carcinogens, and greater retention in the alveoli could cause an increased risk from both lung cancer and cancer in other organs which accumulate these chemicals after their deposition in the lungs.

1.4 Limitations of Particle Size-Selective Sampling and Samplers

The effective application of size-selective sampling procedures to respiratory hazard evaluation requires: a) adequate knowledge of the regional deposition and clearance of particles in people; and b) practical, reliable, reproducible, and accurately calibrated size-selective samplers.

It is apparent from reviews of human deposition and clearance data and models that regional deposition is not fully understood and that the predictive models are at best approximations. Furthermore, there are very large variations in both regional deposition efficiencies and clearance rates among normal people and within individuals at different stages of activity. Thus, even if the data were highly precise and reproducible, the individual risks from the inhalation of a given aerosol could vary over a wide range.

The technology for designing a particle size-selective sampler and characterizing its collection characteristics is more advanced than that for determining regional deposition and clearance dynamics in the human respiratory tract. Yet, even here, there are many conflicting data in the literature, and many instrument designs have required significant modifications to meet their original specifications. Also, much laboratory calibration data have not been matched by instrument performance under field test conditions. Some of these instruments, especially those most applicable to the charge of the Committee, are reviewed in later chapters. With this in mind, the recommendations made in this book reflect the present state of the art, with a full awareness of the need for further research.

In later chapters, the features of particle size-selective sampling systems are reviewed and generic particle size-selective sampling system requirements are described. For toxic particles present in the workplace, the analysis attempts to include size-selection of the particle fractions that can penetrate or deposit and have a biological effect on various regions within the pulmonary system. The resultant practical procedures have been designed with sampler collection characteristics matching the criteria for the inhalable, thoracic and respirable fractions. Since the aim is to approximate the potential dose of the material to the respiratory system, those procedures require the use of integrating mass collection devices (e.g., based on time-weighted average, or TWA, sampling). The strengths and weaknesses of the individual samplers are described and the efficiencies of the collection curves are presented and evaluated in terms of the extent to which they approximate to the shape of the sampling efficiency curve for the appropriate regions of the human respiratory tract. An exception is particle size-selective sampling for diesel exhaust in coal mines, where two different components of the respirable fraction, both composed primarily of elemental carbon, are present in the occupational environment. In this case, the sampling approach is based on the quite different particle size distributions for the two components. A sharp cut-off at $d_{ae} = 0.8$ µm is used, with almost all of the diesel exhaust particles lying below that cut size, and almost all of the coal dust lying above it.

1.5 Chemical composition versus particle size

For ambient air sampling, where the chemical composition of fine particles, predominantly from combustion sources, differs markedly from coarse particles, predominantly generated mechanically from soil or soil-like sources, a sharp cut at d_{ae} = 2.5 μm has been adopted by EPA in order to be able to analyze the concentrations of fine and coarse particles separately. The fine particles appear to generally represent the greatest hazard in terms of excess mortality and morbidity among sensitive segments of the population, and better control of fine particle sources cannot be effectively or efficiently accomplished through control programs using PM_{10} as the primary or sole index of ambient air quality for particulate matter (PM).

For occupational health applications of size-selective sampling, where exposure limits are generally more chemical specific, and the deposition of particles in the respiratory system varies according to the aerodynamic particle size, collection and analysis of size segregated samples by the specified procedures can be used to determine how much of a chemical substance will be available for deposition in the interior regions of the respiratory system and compared to appropriate size-selective TLVs. Unfortunately, as has been shown in many industrial situations, the particles often will not have a homogeneous chemical composition and a substance will not necessarily be distributed uniformly, as a percentage of the mass, over the range of particles found in a particular work setting. Examples for various materials are illustrated in Chapter 10.

The diversity of chemical and physical characteristics of individual particles requires the development of compound and PSS-TLVs. Sampling procedures are identified for use in instances where one or more particle size ranges are expected to contain a particular compound. However, knowledge of the dissolution rate and volatility of each compound is also required in order to evaluate the potential for toxicological impact at various target sites within the body after inhalation and deposition in the respiratory tract. The latter point will necessitate the acquisition of further information for inclusion in the TLV background documentation.

The process of defining a TLV has always dealt with the assimilation of information on the diseases associated with a particular substance. It should now be apparent that for particles, the development of TLVs appropriate for protection of workers requires even more information. The PSS-TLV must include information of the particle size associated with a substance, its effects after deposition, and its rate of dissolution in the various regions of the lung. Chapter 10 reviews the history of the TLVs and describes how toxicological, epidemiological and particle aerodynamics are integrated into effective, protective TLVs.

1.6 Summary

The development of particle size-selective occupational threshold limit values has usually not proceeded in a consistent fashion in the past, either within ACGIH or in standards setting bodies elsewhere. But the development of a reliable data base on size-selective particle deposition in the human respiratory tract in recent years has enabled the establishment of a truly scientific rationale for the specification of new or additional PSS-TLVs. These should consider the diseases associated with the inhaled substance and should be based upon the physical characteristics of the lung, size-mass distribution and dynamics of particles, the physical and chemical composition of particles emitted by varying processes, and other factors including dissolution rates in the lung.

References

ACGIH Technical Committee on Air Sampling Procedures (1984), Particle size-selective sampling in the workplace, *Ann. Am. Conf. Govt. Ind. Hyg.*, 11, 21.

ACGIH-AIHA Aerosol Hazards Evaluation Committee (1975), Recommended procedures for sampling and counting asbestos fibers, *Am. Ind. Hyg. Assoc. J.*, 36, pp. 83-90.

Albert, R.E., Lippmann, M. and Briscoe, W. (1969), The characteristics of bronchial clearance in humans and the effects of cigarette smoking, *Arch. Environ. Health*, 18, pp. 738-755.

American Conference of Governmental Industrial Hygienists (ACGIH) (1968), *Threshold Limit Values for Airborne Contaminants for 1968*, p. 17, ACGIH, Cincinnati, OH.

American Conference of Governmental Industrial Hygienists (ACGIH) (1986), *1986-1987 Threshold Limit Values and Biological Exposure Indices*, ACGIH, Cincinnati, OH.

American Conference of Governmental Industrial Hygienists (ACGIH) (1994), *1994-1995 Threshold Limit Values and Biological Exposure Indices*, ACGIH, Cincinnati, OH.

Bolton, R.E., Vincent, J.H., Jones, A.D. *et al.* (1983), An overload hypothesis for pulmonary clearance of UICC amosite fibres inhaled by rats, *Brit. J. Ind. Med.*, 40, pp. 264-272.

British Medical Research Council (BMRC) (1952), Recommendations of the BMRC panels relating to selective sampling. From the Minutes of a joint meeting of Panels 1, 2, and 3 held on March 4, 1952.

Comité Européen Normalisation (CEN) (1992), *Size Fraction Definitions for Measurement of Airborne Particles in the Workplace*, CEN Standard EN 481. CEN, Brussels.

Fang, C.P. and Lippmann, M. (1995), Development of thoracic inlets for asbestos fiber sampling, Presented at 1995 American Industrial Hygiene Conference, Kansas City, MO.

Gehr, P., Schurch, S., Geiser, M. *et al.* (1990), Retention and clearance mechanisms of inhaled particles, *J. Aerosol Sci.*, 21, Suppl. 1, S491-S496.

Hatch, T.F. and Gross, P. (1964), *Pulmonary Deposition and Retention of Inhaled Aerosols*, Academic Press, New York.

HEI Asbestos Literature Review Panel (1991), *Asbestos in Public and Commercial Buildings*, Health Effects Institute–Asbestos Research, Cambridge, MA.

Horsfield, K., Dart, G., Olson, D.E. *et al.* (1975), Models of the human bronchial tree, *J. Appl. Physiol.*, 31, pp. 83-90.

International Standards Organisation (ISO) (1983), *Air Quality — Particle Size Fraction Definitions for Health Related Sampling*, ISO/TR 7708-1983 (E), ISO, Geneva.

International Standards Organisation (ISO) (1992), *Air Quality — Particle Size Fraction Definitions for Health-Related Sampling*, CD 7708, ISO, Geneva.

Leidel, N.A., Bayer, S.G., Zumwalde, R.D. *et al.* (1979), *USPHS/NIOSH Membrane Filter Method for Evaluating Airborne Asbestos Fibers*, DHEW (NIOSH) Pub. No. 79-127, NIOSH, Rockville, MD (February 1979).

Lippmann, M. (1988), Asbestos exposure indices, *Environ. Res.*, 46, pp. 86-106.

Lippmann, M., Gearhart, J.M. and Schlesinger, R.B. (1987), Basis for a particle size-selective TLV for sulfuric acid aerosols, *Appl. Ind. Hyg.*, 2, pp. 188-199.

Miller, F.J., Gardner, D.E., Graham, J.A. *et al.* (1979), Size considerations for establishing a standard for inhalable particles. *J. Air Pollut. Control Assoc*,; 29: 610-615.

Morrow, P.E. (1988), Possible mechanisms to explain dust overloading of the lungs, *Fundam. Appl. Toxicol.*, 10, pp. 369-384.

Morrow, P.E., Gibb, F.R., and Johnson, L. (1964), Clearance of insoluble dust from the lower respiratory tract, *Health Physics*, 10, pp. 543-555.

National Institute for Occupational Safety and Health (NIOSH) (1975), *Criteria for a Recommended Standard — Occupational Exposure to Cotton Dust*, DHEW (NIOSH) Pub. No. 75-118, U.S. Government Printing Office, Washington. DC.

National Research Council (1980), *Measurement and Control of Respirable Dust in Mines*, NMAB-363, National Academy of Sciences, Washington, DC (1980).

Orenstein, A.J. (Ed.) (1960), *Proceedings of the Pneumoconiosis Conference*, Johannesburgh, 1959. J. and A. Churchill, Ltd., London.

Phalen, R.F. (Ed.) (1985), *Particle Size-Selective Sampling in the Workplace*, Report of the ACGIH Air Sampling Procedures Committee, American Conference of Governmental Industrial Hygienists (ACGIH), Cincinnati, OH.

Public Law 91-173, Federal Coal Mine Health and Safety Act of 1969, 91st Congress (December 10, 1969).

Public Law 91-596, Occupational Safety and Health Act of 1970. 91st Congress (December 29, 1970).

Public Law 95-164, Federal Mine Safety and Health Act of 1977. 95th Congress (November 9, 1977).

Rubow, K.L., Marple, V.A., Tao, Y. *et al.* (1990), Design and evaluation of a personal diesel aerosol sampler for underground coal mines, Preprint No. 90-132. Society for Mining, Metallurgy, and Exploration, Littleton, CO (1990).

Soderholm, S.C. (1989), Proposed international conventions for particle size-selective sampling, *Ann. Occup. Hyg.*, 33, pp. 301-320.

Sutton, G.W. and Reno, S.J. (1968), Respirable mass concentrations equivalent to impinger count data, Presented at American Industrial Hygiene Conference, St. Louis, MO (May 1968).

United States Environmental Protection Agency (EPA) (1996a), *Air Quality Criteria for Particulate Matter*, EPA/600/P-95/001, Washington, DC.

United States Environmental Protection Agency (EPA) (1996b), *Review of the National Ambient Air Quality Standards for Particulate Matter*, Office of Air Quality Planning and Standards (OAQPS) Staff Paper, EPA-452/R-96-013, Environmental Protection Agency, Research Triangle Park, NC (July 1996).

United States Environmental Protection Agency (EPA) (1997), National Ambient Air Quality Standards for Particulate Matter, *Federal Register* (July 1997).

Walton, W.H. (1954), Theory and size classification of airborne dust clouds by elutriation. *Br. J. Appl. Phys.*; 5(Suppl. 3): S29-S40.

Walton, W.H. and Vincent, J.H. (1998), Aerosol instrumentation in occupational hygiene: an historical perspective, *Aerosol Sci. Tech.*, 28, pp. 417-438.

Weibel, E.R. (1963), *Morphometry of the Human Lung*, Academic Press, New York.

Wolff, G.T. (1996), Report of the Clean Air Scientific Advisory Committee (CASAC) Technical Subcommittee for Fine Particle Monitoring (Letter Report), EPA-SAB-CASAC-LTR-96-009, U.S. Environmental Protection Agency, Washington, DC (August 7, 1996).

Chapter 2

AIRWAY ANATOMY AND PHYSIOLOGY

Robert F. Phalen

Air Pollution Health Effects Laboratory, Community and Environmental Medicine, and Center for Occupational and Environmental Health, University of California at Irvine

2.1 INTRODUCTION

Because an appreciation of the quantitative aspects of the deposition patterns, persistence times, and translocation pathways of inhaled materials is important to developing a rational basis for particle size-selective aerosol sampling, it is useful to consider certain relevant aspects of the normal anatomy and physiology of the respiratory tract. Anatomical and physiological information in this chapter is limited to the normal, non-diseased state, which may apply to some members of the smoking population, but may not apply to people with emphysema, bronchitis, asthma, fibrosis, respiratory tract cancer and other diseases. It is also known that variations in normal structure and function occur in the adult population, but because of the paucity of quantitative information, one is limited for the most part to describing the "average" or "typical" adult respiratory tract.

2.2 REGIONS OF THE HUMAN RESPIRATORY TRACT

Following the path of inhaled air, the anatomical structures of the respiratory tract include: 1) the nose, consisting of the nares, vestibule, and nasal cavity proper (with the conchae or turbinates); 2) the nasopharynx; 3) the lips and oral cavity; 4) the oropharynx; 5) the laryngopharynx; 6) the larynx; 7) the trachea; 8) the bronchi; 9) the bronchioles; 10) the respiratory bronchioles; 11) the alveolar ducts; 12) the alveolar sacs; and finally 13) the alveoli. These structures are commonly grouped into two or more separate regions for purposes of simplification. Just how these regions should be identified is not obvious and several compartmentalization schemes are in use.

Two, essentially equivalent, regional models are especially relevant to a consideration of particle size-selective aerosol sampling in the workplace and other environments. The two models are those of the Task Group of the International Commission on Radiological Protection (ICRP, 1994), and the Respiratory Tract Dosimetry Modeling Committee of the National Council on Radiation Protection and Measurements (NCRP, 1997). Table 1.1 (see Chapter 1) displays the major regions of each of these two models along with the ACGIH Air Sampling Procedures Committee's recommended terminology. The respiratory tract is divided into three portions, based both upon anatomical features and upon particle deposition and clearance phenomena that occur within each region. Identification of these three major regions in this way has been of great use to inhalation toxicologists and industrial hygienists.

The usefulness of these three compartments can be illustrated by considering the deposition of inhaled particles. Deposition phenomena for particles can be complex, especially when they pass through an intricate geometrical structure such as the respiratory airways. If one measures the concentration of particles in inhaled and exhaled air and makes a plot of deposition versus particle size, a valley-shaped deposition efficiency curve is obtained (Stöber *et al.*, 1991). The minimum in the total deposition curve at about 0.5 μm occurs because particles of this diameter are not strongly influenced by either inertial or diffusional forces. By using more detailed measurement of regional deposition and performing some mathematical calculations, one is able to break this total deposition curve into three components, one for each region. The deposition probabilities in each region can sometimes be used to relate the anatomical locations of various diseases to the sizes of particles that tend to cause these diseases. For example, nasal cancer in machinists and woodworkers can be related to the high nasal collection efficiency for airborne particles above 10 μm in aerodynamic diameter. Such particles are produced by grinding and sanding operations and by wood cutting.

Region 1, the *head airways region* (HAR), begins at the anterior nares and includes the respiratory airway through the larynx. Particle deposition in this region is primarily limited to large particles whose inertial properties cause impaction in the oral or nasal passages or entrapment by nasal hairs, and to very small particles which are brought into contact with the airway walls by diffusion. Several pathways, with half-times of minutes to days, are used to describe the clearance of particles which deposit in the nasal airways. The rapid half-times describe uptake of relatively soluble material into the blood, and physical clearance by mucociliary transport to the throat for subsequent swallowing. The slow half-times represent binding of particle compounds to tissues, or deposition on poorly-cleared surfaces. Experimental data indicate that the anterior one-third of the nose, where about 80% of particles with aerodynamic diameter 7 μm deposit, does not clear insoluble particles except by blowing, wiping or other extrinsic means.

Region 2, the *tracheobronchial region* (TBR), begins at the trachea and includes the ciliated bronchial airways down to and including the terminal bronchioles. A relatively small fraction of particles of all sizes which pass through the HAR will deposit in the TBR. The mechanisms of inertial impaction at airway bifurcations, sedimentation and, for small particles, Brownian diffusion are the ones primarily responsible for TBR deposition. But interception too can be an important deposition mechanism for fibrous dusts. During mouth breathing, the benefits of the collection of larger particles in the nose are largely lost and these larger particles tend to deposit in the TBR with high efficiency. An important characteristic of TBR is that it is both ciliated and equipped with mucus secreting elements so that clearance of most deposited particles occurs probably within 24 hours by mucociliary action to the throat for swallowing. However, there is evidence to suggest that there is a slow-clearing component of the TBR (ICRP, 1994). Again, relatively soluble material may quickly enter the blood stream. The rate of mucus movement is slowest in the finer airways and increases toward the trachea (NCRP, 1997). Clearance of material in this region cannot be described by a single rate. Depending on the location of deposition and particle properties, such as size and solubility, clearance half-times range from minutes to tens of days (Morrow *et al.*, 1967; Stöber *et al.*, 1991; ICRP, 1994; NCRP, 1997).

Region 3, the *gas exchange region* (GER), includes the functional gas exchange sites of the lung. It includes respiratory bronchioles (RB), alveolar ducts, alveolar sacs, and alveoli. For particles to reach and deposit in this region, they must penetrate the two more proximal regions on inspiration and by settling, diffusion, or interception come into contact with deep lung surfaces. Since there is gas exchange between tidal and residual air, a portion of each breath remains un-exhaled and the times available for deposition may be long for some particles. Clearance from this region is still not completely understood, but several mechanisms are believed to exist including: a) the dissolution of relatively soluble material with absorption into the systemic circulation; b) direct passage of particles into the blood; c) phagocytosis of particles by macrophages with translocation to the ciliated airways; and d) transfer of particles to lymphatic channels, vessels and lymph nodes (Morrow *et al.*, 1967; Stöber *et al.*, 1991; ICRP, 1994; NCRP, 1997). Insoluble particles which are transferred to the lymphatic system are probably not cleared. Further, a proportion of insoluble particles depositing in the GER becomes "sequestrated" (Vincent, 1995), either in immotile macrophages or in fixed tissue, and those particles too are not cleared and so, over a long period of time, contribute to the cumulative lung burden.

The 3-region model does have some rather important drawbacks. For example, the pattern with which particles deposit within a given region is not usually addressed. The assumption that deposition within a given region is uniform may eventually lead to improper estimations of risk. For example, bifurcations in the tracheobronchial region can be sites of high regional particle deposition. Also, the model does not adequately separate out the region between the terminal bronchioles and the alveolar ducts and sacs. This junctional region contains respiratory bronchioles (RB) that are unique in structure in that they have both air conducting and gas exchange properties. Here, deposition of inhaled particles appears to be greater than in more distal regions, presumably due to lack of convective penetration of air beyond the RB. Because this portion of the gas exchange region is often the site of airway disease in humans, it should not be overlooked.

2.3 Gross Anatomy

Nose, Nasopharynx, Oral Cavity, Oropharynx and Larynx

The mammalian nose and its immediately postnasal cavities comprise an elaborate organ that provides for olfaction, detection of airborne irritants, collection of noxious gases and particles, humidification and temperature adjustment of inspired air, and disposal of fluids that drain from the eyes, sinuses, and inner ears. The importance of these functions to maintaining good health also makes the nasal-pharyngeal region an important target for airborne agents. Despite this fact, the nasal region is often overlooked in industrial hygiene. But it must frequently deal with air pollutants in their raw unfiltered state at ambient concentrations, and failure of any of its critical functions can lead to serious, even life threatening conditions.

In humans, the nose contains two channel-shaped nasal cavities that are separated by a cartilaginous and bony septum. The average adult male's nose has an air volume of about 17 ml. Each nasal cavity is entered through a *naris* (nostril) having a cross-sectional area of about 0.7 cm^2 (Landahl, 1950; ICRP, 1975). The nasal cavity is supported by walls consisting of bone, cartilage, and connective tissue that provide sufficient rigidity to prevent collapse during breathing. The *anterior* (nearest to the nares) one-third of the nasal cavity is covered with skin much like that on the face and does not have an effective coating of mucus. The *posterior* (rearward) two-thirds of the cavity is covered with mucus that moves rearward driven by cilia at an average velocity of about 1 cm per minute to a point where it is swallowed. This mucus, produced by goblet cells and glands, is mixed with fluids, including tears, that drain into the nasal cavity from the eyes and sinus cavities of the facial bones. The olfactory region has specialized cells that are covered with mucus, but have no motile cilia. The anterior portion of the adult's nasal cavity is partially covered with hair that traps large inhaled bodies and warns of their presence via nerves at the base of the hair follicles.

At its rear, the nasal cavity narrows and turns sharply downward. This area, the *nasopharynx*, is a region for the impaction of large particles that have eluded previous capture. The nasopharynx, roughly tubular in shape, is joined by the oral pharynx (rear portion of the mouth) a few centimeters down its length. The pharynx then divides at the epiglottis to turn and enter either the larynx and trachea or, continuing downward, the esophagus. The pharynx is coated with mucus in the same fashion as the nasal cavity.

The mouth, entered through the variable-sized opening between the lips, is divided into two regions, an anterior vestibule or labial cavity (including the inner lips, cheeks and teeth) and a posterior or buccal cavity which joins the oropharynx. The oral cavity is normally about 70 to 75 mm in length (antero-posterior dimension), 40 to 45 mm in horizontal dimension, and 20 to 25 mm in vertical dimension. The tongue has a volume of approximately 60 to 70 cm^3 (ICRP, 1975). The shape and cross-sectional area of the lip opening and oral cavity are highly variable, in response to oxygen demand and vocalization.

The *epiglottis* is a muscular flap that moves to cover the entrance to the larynx and trachea during swallowing. Other muscular action also prevents swallowed material from entering the trachea because persons whose epiglottis has been surgically removed can still swallow without choking.

The *larynx*, or voicebox, is a short tubular cavity that has a slit-like variable-sized narrowing in its central portion. The narrowing is caused by two pairs of folds in the walls of the larynx. The uppermost folds are called the false vocal chords and the lower folds the true vocal chords. The adult larynx is about 3.5 to 5 cm long and has a variable cross-section that depends on the air flow rate passing through it (Stanescu *et al.*, 1972). The larynx represents a major resistive element to air flow and also forms an inspiratory air jet that leads to particle impaction on the rear wall of the trachea (Schlesinger and Lippmann, 1976). The larynx is encased by muscle, bone, and cartilage, and it is lined by a mucus-covered membrane very similar to that found in the rear portion of the nasal cavities and pharynx. In the larynx, mucus is propelled upward for swallowing.

TRACHEOBRONCHIAL TREE

Trachea: At its entry the *trachea* is continuous with the larynx. It is a flexible tube that divides into primary bronchi. In humans, about 20 roughly U-shaped cartilage rings in its wall prevent tracheal collapse. The gap between the ends of each cartilaginous ring is covered by a flexible muscular sheet of tissue. Thus, in cross section, the trachea tends to have a D- or O-shape depending upon the internal air pressure and hence the air flow rate. The inner walls of the trachea are covered with mucus, which is supplied by goblet cells and mucous glands. During breathing, both the diameter and length of the trachea vary, the trachea elongating on inspiration.

Bronchi and bronchioles: In humans, the trachea divides into two main branches called "*major bronchi*." These bronchi enter the right and left lungs and continue to divide for several generations, averaging about 16 in humans, before *alveoli* (air sacs) begin to appear in the bronchiolar walls as openings into the lumen. This appearance of alveoli marks the end of the tracheobronchial tree and the beginning of the GER. The *bronchi* and *bronchioles* are roughly circular in cross section, and smooth muscle, which is present only along one side of the trachea, completely encircles the bronchial airways. The U-shaped cartilages of the trachea are replaced in the bronchial walls by irregularly-shaped cartilage plates situated outside of the smooth muscle. Further down the tracheobronchial tree where the tube diameters are about 1 mm or less, the cartilage disappears. These tubes are the bronchioles. Bronchioles have mucus-secreting goblet cells but do not have mucous glands in their walls. The outermost layer of the bronchi consists of a mixture of tough connective tissue and elastic fibers. The inner lining of the bronchi is pseudostratified columnar epithelium having ciliated cells, mucus-secreting goblet cells, and underlying mucus-secreting glands. Thus the tracheobronchial tree possesses an active clearance mechanism due to the propulsion of mucus toward the pharynx to a point where it is swallowed. The bronchioles are lined with ciliated columnar epithelium that is not pseudostratified. The symmetrical tracheobronchial tree model for the human, described by Weibel (1963), is widely used and contains information on airway lengths, diameters and number. Similar, but asymmetric, human models have been published by Horsfield and Cumming (1968) and by Yeh and Schum (1980).

RESPIRATORY BRONCHIOLES (RB)

In humans, the *terminal bronchioles* of the tracheobronchial tree (i.e., those with diameters of about 0.6 mm) branch to form the first-order *respiratory bronchioles* (RB). These RB continue to divide and branch to give a total of about 2 to 5 orders of tubes. RB, as they branch, exhibit an increasing number of alveoli opening into their lumens. These alveoli are thin-walled, surrounded by blood capillaries, and presumably participate in the gas exchange function of the lung. Because of this gas exchange, these bronchioles are called respiratory. Between alveoli, the surface of the RB is thicker epithelium. Within alveoli, ciliated cells are not found and clearance of deposited debris is presumably by those mechanisms associated with deeper situated alveoli.

Two major points must be made with respect to RB. First, the RB have been acknowledged as an important site for disease in humans. Second, these structures form part of the "silent zone" of the lung, a region in which respiratory disease is very difficult to detect by conventional pulmonary function testing.

PARENCHYMA

"*Parenchyma*" is a term that relates to the primary functional tissue of an organ as distinct from its supporting framework or secondary tissues. When applied to the lung, the parenchyma relates to the alveoli and does not include the trachea and bronchial tree, which are often viewed as merely conductive airways for the purpose of delivering air to and from the gas exchange region. The major structural elements of the parenchyma of the lung are alveolar ducts, alveolar sacs, alveoli, alveolar capillaries and the pulmonary lymphatics.

The *alveolar duct* is a tubular structure whose walls are completely covered with alveoli. The alveolar duct usually branches to either two other alveolar ducts or two blind-ended tubes called alveolar sacs. With respect to the total number of ducts and sacs, Weibel's figures of 7×10^6 ducts and 8.4×10^6 sacs are probably reliable. The pulmonary acinus, which consists of a terminal bronchiole and the structures supplied by it, has been described by Schreider and Raabe (1981).

Although often depicted as spherical, the *alveolus*, or air sac, more closely resembles an incomplete polyhedron. The open face of the alveolus is exposed to the air in either a respiratory bronchiole, an alveolar duct or an alveolar sac, the closed portions being surrounded by a network of fine blood capillaries. In the alveolus, the atmosphere and the blood are brought into intimate contact where equilibration of CO_2 and O_2 can take place. In addition to the surrounding capillary net, alveoli are partially surrounded by elastic and non-elastic fibers that provide mechanical support. Alveoli, capillaries and fibers are embedded in an interstitium, or connective tissue. The average diameter of an adult's 300 million alveoli is about 200 to 300 μm.

2.4 Cells and Tissues of the Respiratory Tract

Ciliated Mucosa

The tissue that lines the rear portions of the nose, the larynx, the trachea, the bronchi and the bronchioles is described as "*ciliated mucosa*." Such tissue is characterized by the presence of cells with numerous tiny hair-like projections (cilia) and by the presence of individual cells and glands that secrete the components that make up mucus — a sticky, viscoelastic fluid. The cilia are motile and beat in a coordinated fashion resulting in movement of the overlying mucus, usually toward the region where the mucus is swallowed.

The ciliated cells of the human respiratory system have cell nuclei and are columnar in shape, being about 10–15 μm in diameter and 20–40 μm in height. The ciliated cells are attached at their bases to a basement membrane and new replacement cells appear to form beneath the mature cells and grow upward to replace cells that are lost. At the top surface, protruding into the lumen of the airway, there are 15 to perhaps 100 or more filamentous cilia that are 5–15 μm long and about 0.2 μm in diameter. The cilia bend and then lash forward at rates up to several hundred cycles per minute. It is the coordinated beating of cilia on adjacent cells that propels the mucus at rates on the order of several millimeters per minute.

Interspersed among the ciliated cells are columnar *goblet cells*, similar in size to the ciliated cells but lacking in cilia and having a narrow base, and thus a goblet shape. These cells, also attached to the basement membrane, manufacture mucus and, when filled, open at the top and discharge their contents onto the airway surface. Beneath the basement membrane, there are mucous glands consisting of clusters of mucus-secreting cells that secrete into a duct that leads to the epithelial surface. The action of the ciliated mucus-secreting tissues is responsible for sweeping surfaces of the airways free of particulate contamination. This function depends upon the quality and quantity of mucus and the quantity and synchronization of cilia. Viral and bacterial infections as well as various toxicants can lead to over- or under-secretion of mucus and to loss or paralysis of cilia. During such states, it appears that sneezing and coughing become the major clearance mechanisms that serve to clear the mucociliary epithelium. Often thought of as an annoying symptom, coughing can be a health-preserving mechanism for removing mucus, toxicants, and infectious organisms from the respiratory tract.

THE ALVEOLUS

As was previously mentioned, the adult human's alveolus is a polyhedral structure, about 200–300 μm in diameter, having one face open to the airway. The walls of this structure are formed by very thin alveolar epithelial cells whose nuclei may at times bulge into the alveolar airspace. In reality there is more than one type of alveolar epithelial cell. At its thinnest portions, the type I (also called type A) alveolar epithelial cell is about 0.1 μm or slightly less in thickness (Schreider and Raabe, 1981). These cells appear to have relatively smooth surfaces and lie on top of a basement membrane that is about 0.02 to 0.04 μm thick. Another basement membrane supports the blood capillary endothelial cells. These endothelial cells join to form the capillary wall and are quite similar in size and shape to the thin alveolar cell. The total thickness of the air to blood interface has been measured by Meessen (1960) and reported by Weibel (1963) and Weibel and Gill (1977) to be between 0.36 and 2.5 μm.

A thicker, roughly cube-shaped cell, the type II (or type B) epithelial cell of the alveolus, has a surface covered with small protrusions (microvilli). These microvilli on the airspace side greatly increase the surface area of this cell and imply, along with the presence of inclusions within the cell body, that this cell manufactures and secretes substances onto the surface of the alveolus. Biochemical and other evidence indicates that this cell is involved in the manufacture and secretion of surfactant, a surface tension-lowering agent that reduces the tendency alveoli have for collapsing (Weibel and Gill, 1977; Parent, 1991). Abnormalities in lung surfactant can be related to a variety of disease states. The type II cell, which is capable of mitotic division, may serve as a precursor (parent) to the type I cell, both during lung growth and repair of injury.

Other cells present in the alveolar region include *macrophages*, *alveolar brush cells* and *interstitial cells*. The alveolar brush cell which is described as roughly pyramidal, sits on the alveolar basement membrane and protrudes into the alveolar lumen. It has large microvilli on its mucous-covered side and has as yet unknown functions. But several functions have been proposed including regulation of depth or viscosity of mucus, and chemosensory.

In some areas, the basement membranes of the alveolus and capillaries are separated by a space called the "*inter-alveolar septum*" or "*interstitium*." This interstitium contains both elastic and inelastic fibers and cells called "*fibroblasts*." Fibroblasts are irregularly-shaped cells that are involved in the formation of connective tissue. Rarely, nerve fibers have been seen in the interstitium (Nagaishi, 1972). In pathologic conditions such as edema (fluid accumulation) and infection, the interstitial space may become enlarged due to the presence of excess fluid and cells such as blood leukocytes (white cells).

Alveolar walls are frequently observed to have pores that appear to connect the airspaces of adjacent alveoli. These pores, called "*pores of Kohn,*" were discovered by Adriani in 1847 according to Miller (1947). Little information is available on their shapes, numbers, or dimensions.

THE MACROPHAGE

Alveolar macrophages are relatively large (their average volume in the human lung is 2500 μm^3; Crapo *et al.*, 1982), nucleated cells which possess the ability to engulf foreign materials. Roughly similar to the familiar amoeba, macrophages can change shape presumably by: a) liquefaction of their cell membrane; b) subsequent flowing of the cell contents; and c) reformation of their surface membrane. Thus, the cells are mobile and can move to surround and engulf foreign material in their paths.

"*Phagocytosis*" and "*pinocytosis*" are two terms used to describe the engulfment of substances in varying states by cells, such as macrophages. Phagocytosis refers to the incorporation of solid materials, while pinocytosis refers to the incorporation of liquid droplets. A third term, "*endocytosis,*" includes both phagocytosis and pinocytosis. Macrophages are found on the surfaces of the alveoli in the deep lung and are not a fixed part of the alveolar epithelial wall. They are credited with maintaining the sterility of the lung by virtue of their ability to engulf and kill infectious microorganisms such as bacteria. Macrophages also engulf dusts and other fine particles that deposit in the deep lung upon inhalation. It appears that these pulmonary alveolar macrophages (PAM) undergo chemotaxis, i.e., movement in response to chemical stimuli. Chemotaxis may be positive, toward the stimulus, or negative, away from it.

The process of phagocytosis has been described as occurring in seven sequential steps by Stossel (1976). The steps are: a) target recognition; b) reception of the message to initiate phagocytosis; c) transmission of the message to an effector; d) attachment of the macrophage membrane to the target; e) formation of pseudopodia; f) engulfment by the pseudopodia; and g) fusion of the pseudopodia with the macrophage cell body. Failure of any of these phagocytic subprocesses could possibly result in the inactivation of the function of the macrophage in providing for defense of the lung.

Hocking and Golde (1979) have reviewed the research on antimicrobial mechanisms of the macrophage and listed several chemicals, present within these cells, that have microbicidal properties. Included in the list were hydrogen peroxide, catalase, superoxide anion, and lysosomal cationic proteins. With respect to nonviable material deposited in the deep lung, Brain and Corkery (1977) have suggested that a major role of the macrophage is the prevention of incorporation of materials that have slow rates of clearance into tissues of the lung. For example, the alveolar epithelium and alveolar interstitial spaces appear to have quite slow turnover times for particulate contaminants. Thus, if phagocytosis occurs, potentially toxic particles are prevented from entering a tissue compartment in which they will be in contact with vulnerable cells for protracted periods of time.

Macrophages have an amazing efficiency for engulfing particles. Within minutes of deposition of an inhaled particle, the pulmonary alveolar macrophage is seen to have begun ingestion. Also, these cells appear to be able to phagocytize even when packed nearly full of debris. On the other hand, certain dusts are clearly toxic to the macrophage and result in their death or debilitation. Examples of such cytotoxic materials include particles of cadmium, nickel, manganese, chromium, silica, vanadium, plutonium, and coal dust (Hocking and Golde, 1979; Brain and Corkery, 1977). Macrophages appear to efficiently engulf a relatively narrow size range of particles (Fenn, 1921 and 1923). Holma (1969) suggested 1.5 μm optimal particle diameter of spherical particles for maximally efficient uptake by macrophages. He found that phagocytic uptake had an upper limit on particle diameter of 8 μm, a size which usually does not penetrate to alveoli. Fibers are exceptions in that alveolar deposition occurs for particles whose lengths exceed the limits for phagocytosis.

MUCUS-SECRETING GLANDS

Mucus-secreting glands are present in the nose and tracheobronchial tree. These glands are present in greater numbers in large airways and become more sparse moving down to smaller airways, disappearing at the level of the bronchiole. Along with goblet cells, these glands produce the mucus that covers the ciliated portions of the respiratory tract. Because these glands lie beneath the mucous membrane, they are called "*submucosal glands.*" They have a branched tubuloacinar structure (where "*acinar*" refers to the blind ends of the tubes that branch and form each gland). The tubes into which mucus is secreted join a collecting duct that becomes ciliated just before it enters the bronchial airspace. These ciliated ducts appear as many pinholes on the surfaces of bronchi, having a maximum surface concentration of about one opening per mm^2 in the trachea (Netter, 1979).

Two types of cells, mucus and serous, rest on a basement membrane and line the tubules. *Serous cells* are found lining the blind ends of the tubules, and *mucus* cells line the more proximal (upper) portions. The secretions of these cells, mucus, is primarily an acid glycoprotein which is both viscous and elastic.

INNERVATION OF THE RESPIRATORY SYSTEM

The nervous system receives, generates, conveys, stores, and processes information. Portions of the nervous system, found in nearly every tissue of the body, play an important part in the voluntary and involuntary control and coordination of muscles, organs, glands and their sub units, tissues and cells. In the respiratory system, nerves are responsible for: a) control of muscles for breathing, adjustment of the size of bronchial airways, and control of the cough, sneeze and gag reflexes; b) the initiation and control of protective breathing patterns; c) the control of secretions; d) adjustment of the distribution of blood flow; and e) provision of sensory information on odor, irritancy, and the composition of lung tissue fluids and blood. As for the body in general, much of the information that is carried by the nervous systems of the respiratory tract is not noticed at the conscious level.

Especially important are nerves that trigger the cough reflex; nerves that lead from pressure, stretch and chemical receptors; and nerves involved in bronchial muscle constriction, protective breathing patterns, and mucous gland secretion. It is clear that the innervation of the respiratory tract is extensive and, in fact, present in nearly every region from the nose down to the alveoli. The interaction of inhaled airborne toxicants with this system is not well understood at present.

BRONCHIAL MUSCULATURE

Smooth muscle that runs along the rear wall of the trachea and forms a spiral wrap around bronchi and bronchioles can greatly reduce the airway diameter via constriction. Constriction of bronchial muscles is a major precipitating factor, along with excess mucus secretion, in an asthma attack. Such an attack can be caused by inhaled particles or gases, infection, ingested foods or drugs, and psychological factors. Bronchial constriction can be due to direct chemical action on smooth muscle, via a nerve reflex or by liberation of histamine and other chemicals by cells.

2.5 BREATHING

NORMAL BREATHING

Under resting conditions, the average adult male inhales about 500 cm^3 of air per breath (the tidal volume). This volume is small in comparison to a total lung capacity of about 6 liters. The inspired air fills the nose and tracheobronchial tree which has a cumulative total volume or about 200 cm^3. Alveolar ventilation as measured in the laboratory is about 300 cm^3 of the inspired 500 cm^3 of air. This 300 cm^3 mixes with the air already residing in the GER. Upon normal exhalation, the air that was in the tracheobronchial tree from the previous inspiration is essentially completely expelled from the body along with about 300 cm^3 of alveolar air. But this exhaled alveolar air is not the same 300 cm^3 that was just inhaled; rather it is 300 cm^3 of mixed GER air. Thus, pollutant gases or particles that are inhaled into the GER may be washed out over several successive breaths. The times available for deposition and uptake of inhaled materials in the bronchial tree are short (on the order of a few seconds), while residence times for deposition in the GER can be on the order of tens of seconds or even minutes.

During quiescent breathing (*eupnea*), inhalation begins about every 4–5 seconds and brings the volume of gas in the lungs up to about half of the maximum or total lung capacity. Exhalation reduces the volume of gas in the lung by about 500 cm^3, leaving the lung at its resting expiratory level or functional residual capacity with a gas volume of about 2500 cm^3.

EXERCISE

The effects of exercise on the magnitude of deposition of airborne material is of critical importance in occupational and environmental hygiene considerations. During exercise, several events occur that may change the response to a toxicant. The increased volume of inhaled air leads to greater tissue exposure. A person walking may inhale 2–3 times more air per minute than while sitting, and during maximal exertion the minute volume can be increased more than 10-fold. An offsetting factor to the greater rate of delivery of an airborne toxicant during exercise is the fact that the functional residual capacity may be increased in the exercise state. Thus the freshly inhaled material is diluted by a larger volume of residual air. Usually, the percent increase in functional residual capacity will be small with respect to the percent increase in minute ventilation. In exercise there is often a shift from nasal to oronasal (combined nose and mouth) breathing. This shift, which occurs at different work loads in different subjects (Saibene *et al.*, 1978), has the effect of decreasing the resistance of the upper airways to air flow. The fraction of air that enters the mouth does not receive the benefit of nasal capture of particles and pollutant gases. This bypass of an important respiratory tract defense will tend to increase the biological impact of an inhaled toxicant.

Other factors in exercise that may modify the subject's response to an air pollutant are: a) the potential overriding of protective reflex breathing pattern changes (e.g., a protective shift to more shallow breathing); b) the widening of the larynx in conjunction with increased flow; and c) changes in tissue metabolism which may alter local tissue sensitivity.

2.6 REFLEX RESPONSES TO INHALED IRRITANTS

The rich innervation of the mammalian respiratory tract produces an important category of interaction with inhaled pollutants; viz. reflex changes in breathing pattern due to sensory irritation. In general, these reflex actions appear to be protective in that the changes in breathing pattern serve to limit or prevent exposure of the respiratory tract. Familiar reflex actions include coughing and sneezing. Less familiar are bronchial constriction, reductions in the volumes and rates of ventilation, and excess secretion of mucus. Alarie (1973) has described a classification of inhaled materials, based on their irritancy and effects on various receptors in the respiratory system. The classification is presented in Table 2.1.

Inhaled materials can also produce intoxication, olfaction or pain. Such events, whether pleasurable, neutral or unpleasant can modify behavior, producing avoidance or in extreme cases addiction. A brief review of these aspects of inhaled materials has been published by Wood (1978). The responses to substances with stimulus properties can modify the behavior of people during exposure. Such modifications may act to either produce a greater or alternatively, lesser degree of exposure than would otherwise occur.

2.7 Respiratory Tract Defenses

Bulk clearance mechanisms

The anterior portion of the nose is cleared of deposited insoluble particles by blowing, wiping or other similar extrinsic means; soluble particles may be dissolved and absorbed within tissues or carried away in the blood.

The posterior portions of the nose, where *mucociliary clearance* occurs, show great variability in rates of particle clearance. The variability is due both to the specific location of deposition within an individual and to differences among individuals.

Other mechanisms that operate to rapidly clear large quantities of mucus and entrapped particles include sneezing and coughing. These mechanisms, triggered by physical or chemical irritation, act sporadically and serve as an important backup when mucociliary clearance is impaired or overloaded. Failure of the mucociliary clearance mechanism occurs under several conditions such as excessive thickening or drying of mucus, loss of cilia as occurs during respiratory infections, and inactivation of cilia as occurs during inhalation of toxic gases or smokes. Mossberg (1980) described the role of cough in clearing particles deposited in the tracheobronchial tree. He pointed out that healthy human subjects with normal mucociliary transport were inefficient in clearing 6 μm particles via coughing, but that subjects with impaired mucociliary clearance were usually quite efficient in clearing by cough. His conclusion was that excess quantities of mucus were probably necessary for effective particle clearance during coughing.

TABLE 2.1. Classification of airborne chemicals that stimulate respiratory tract nerve endings*

A. Sensory Irritant
Definition: When inhaled via nose will stimulate sensory trigeminal nerve endings, evoke burning sensation in nose and inhibit respiration. Most will induce coughing from laryngeal stimulation.
Other characteristics: Stimulate trigeminal endings in cornea and induce tearing, induce burning sensation on face, may induce bronchoconstriction.
Examples: Chloracetophenone, SO_2, ammonia, acrolein, inert dust.

B. Pulmonary Irritant
Definition: When inhaled will stimulate sensory receptors in lung and induce rapid shallow breathing. Cause sensation of dyspnea and breathlessness rather than pain.
Other characteristics: Can induce pulmonary edema and then painful breathing.
Examples: Phosgene, NO_2, O_3, sulfuric acid mist, sulfur and nitrogen mustard, sulfur pentafluoride.

C. Bronchoconstrictor
Definition: When inhaled will induce increased resistance to airflow. Action can be directly on smooth muscle, via axonal reflex or by liberation of histamine.
Other characteristics: Most produce pain via action on bronchial mucosa.
Examples: SO_2, ammonia, inert particles, allergens.

D. Respiratory Irritant
Definition: When inhaled acts as sensory irritant, bronchoconstrictor and pulmonary irritant.
Other characteristics: Similar to sensory and pulmonary irritants.
Examples: Chlorine, ketone, dichloromethyl ether, chlorine pentafluoride.

*From Alarie, p. 102 (1973).

He also indicated that cough is most effective in clearing larger airways, but that during successive coughs without intervening inspiration, clearance of smaller airways occurs.

MUCOCILIARY CLEARANCE

In healthy mammals, the tracheobronchial tree is effectively cleared of debris by a moving layer of mucus that is driven toward the oral pharynx and subsequently swallowed. It is generally accepted that the mucus lining is relatively continuous and constantly moving due to the action of ciliated cells, but Van As and Webster (1972) have argued that mucus is transported as plaques, flakes and droplets that do not completely cover the airway surface. In any event, the transport of particles deposited in the tracheobronchial tree appears to proceed in a relatively continuous manner. The completeness of such clearance was challenged by Patrick and Stirling (1977), who examined the clearance of particles injected intratracheally into rats and concluded that approximately 1% of the material remained in the bronchial tissue for at least 30 days. The rates of clearance of particles from the

TBR can be prolonged in disease, and most likely in perfectly healthy individuals. Thus, the ICRP Task Group includes a slow-clearing compartment in the TBR (ICRP, 1994).

The rate of mucus-mediated clearance appears to be faster in larger airways than in smaller ones (Morrow *et al.*, 1967; Stöber *et al.*, 1991; ICRP, 1994; NCRP, 1997). The common use of multiple-clearance half-times to model clearance does not mean that mucus velocities are discontinuous. Such models are mathematically convenient, especially when more than one mechanism is operating. The effects of a variety of factors on particle clearance rates has been reviewed by Pavia *et al.* (1980). Their conclusions are summarized in Table 2.2.

TABLE 2.2. Factors that are either known or believed to influence the clearance of inhaled particles*

Factor	Probable Effect on Clearance Rate
Sex	Males and females probably have similar clearance rates.
Age	Increasing age is associated with slower clearance.
Posture	No apparent effect.
Exercise	Brisk exercise may accelerate clearance.
Sleep	May decrease clearance.
Smoking	Chronic cigarette smoking impairs clearance. Acute exposure to cigarette smoke appears to speed clearance from deep airways and inhibit clearance from large airways. Effects may vary with dose.
Sulfur dioxide	Variable effects seen by different investigators.
Freon propellants	No effect.
Hair spray	Inhibition of clearance.
Chronic bronchitis	Decrease in clearance rate when cough is absent. However more proximal deposition of inhaled particles may mask this effect.
Emphysema	Normal or possibly increased rate of clearance.
Bronchial asthma	Decreased rate of clearance that can, in some cases, be overcome by medication.
Influenza	Decreased clearance with impairment about 2 to 3 months.
Pneumonia	Decreased clearance with impairment lasting up to 1 year.
Bronchogenic carcinoma	Probably no effect.
Cystic fibrosis	Impaired clearance.
Kartegener's syndrome (immotile spermatozoa)	Greatly impaired clearance.
Asbestosis	Probably no effect.

*From Pavia *et al.* (1980).

Other Clearance Mechanisms

The true gas exchange region of the lung does not possess a mucociliary clearance apparatus (respiratory bronchioles may be an exception). Deposited particles are believed to be cleared by several mechanisms including: a) dissolution and uptake by the systemic blood; b) direct passage of very small particles into the blood; c) phagocytosis by macrophages followed by transport to ciliated airways, lyphatics and possibly blood; and d) movement of bare particles into lymphatics (Morrow *et al.*, 1967; Morrow *et al.*, 1977; NCRP, 1997). In addition, particles or their components may remain in the alveolar tissues for very long periods. Inactivation of microorganisms with or without clearance by alveolar macrophages is another important aspect of alveolar defense.

The fate of particles deposited in the GER is strongly dependent on the mechanical stability of the particles. Particles that undergo significant dissolution in the fluids found in the lung may dissolve while still within the air space, inside phagocytes, or while in interstitial spaces. When the total deposited mass of particles is large, insoluble particles may clear at slower than normal rates. This effect, called "*lung overload*" is a topic currently undergoing active research (Bolton *et al.*, 1983; Morrow, 1992). The effect has severe consequences in laboratory rat studies, but its significance in humans is as yet unclear.

2.8 Population Variation

There are inherent anatomical and physiological differences among people associated with differences in body size, gender, and other factors such as state of health. But it is useful to have representative values for normal respiratory parameters and related anatomical features that may be employed for exposure and deposition predictions for inhalation of potentially toxic substances. Such representative values can provide the basis for general comparisons of exposure conditions and expected regional deposition, although it is understood that these values may not describe any particular person. For the purposes of particle size-selective aerosol sampling, such reference values as have been published (ICRP, 1975) can be taken as representative of a 'typical' person (see Table 2.3). However, airway growth is a complex and perhaps individual phenomenon, which has important implications to particle deposition (Phalen *et al.*, 1985). Both the ICRP and NCRP Task Groups have provided age-dependent aerosol deposition models that account for changes in airway size and ventilation with age.

TABLE 2.3. Reference worker: representative values for respiratory anatomical and physiological parameters describing an average healthy worker under light to moderate physical activity

Parameter	Value
Weight	70 kg
Height	175 cm
Age	20–30 years
Body surface area	1.8 m^2
Lung weight	1000 g
Lung surface area	80 m^2
Trachea weight	10 g
Trachea length	12 cm
Total lung capacity	5.6 liters
Functional residual capacity	2.2 liters
Vital capacity	4.3 liters
Residual volume	1.3 liters
Respiratory dead space	160 ml
Breathing rate	15 breaths/min (BPM)
Tidal volume	1450 ml
Inspiratory flow rate	43.5 liters/min
Minute volume	21.75 liters
Inspiratory period	2 seconds
Expiratory period	2 seconds

REFERENCES

Alarie, Y. (1973), Sensory irritation by airborne chemicals, *Crit. Rev. Toxicol.*, 2, pp. 299-363.

Bolton, R.E., Vincent, J.H., Jones, A.D. *et al.* (1983), An overload hypothesis for pulmonary clearance of UICC amosite fibers inhaled by rats, *Br. J. Ind. Med.*, 40, pp. 264-272.

Brain, J.D. and Corkery, G.C. (1977), The effect of increased particles on the endocytosis of radiocolloids by pulmonary macrophages in vivo: competitive and toxic effects, In: *Inhaled Particles IV* (W.H.Walton, Ed.), pp. 551-564, Pergamon Press, Oxford.

Crapo, J.D., Barry, B.E., Gehr, P. *et al.* (1982), Cell numbers and cell characteristics of the normal human lung, *Am. Rev. Respir. Dis.*, 126, pp. 332-337.

Fenn, W.O. (1921), The phagocytosis of solid particles. I. Quartz, *J. Gen. Physiol.*, 3, pp. 575-593.

Fenn, W.O. (1923), The phagocytosis of solid particles. IV. Carbon and quartz in solutions of varying acidity, *J. Gen. Physiol.*, 5, pp. 311-325.

Hocking, W.G. and Golde, W. (1979), The pulmonary alveolar macrophage, *N. Eng. J. Med.*, 301, pp. 580–587 and pp. 639-645.

Holma, B. (1969), The acute effects of cigarette smoke on the initial course of lung clearance in rabbits, *Arch. Environ. Health*, 18, pp. 171-173.

Horsfield, K. and Cumming, G. (1968), Morphology of the bronchial tree in man, *J. Appl. Physiol.*, 24, pp. 373-838.

International Commission on Radiological Protection (ICRP) (1975), *Reference Man,* ICRP Publication 23, pp. 122-124, Pergamon Press, Elmsford, NY.

International Commission on Radiological Protection (ICRP) (1994), *Human Respiratory Tract Model for Radiological Protection*, ICRP Publication 66, Elsevier, Tarrytown, NY.

Landahl, H.D. (1950), On the removal of airborne droplets by the human respiratory tract. I. The lung, *Bull. Math. Biophys.*, 12, pp. 43-56.

Meessen, H. (1960), Die pathomorphologie der diffusion und perfusion, *Verhandl. Deut. Ges. Pathol.*, 44, p. 98.

Miller, W.S. (1947), *The Lung*, Charles C. Thomas, Springfield, IL.

Morrow, P.E. (1977), Clearance kinetics of inhaled particles, In: *Respiratory Defense Mechanisms* (J.D. Brain, D.F. Proctor and L.M. Reid, Eds.), Chapter 14, Part II, pp. 491-543. Marcel Dekker, Inc., New York.

Morrow, P.E. (1992), Dust overloading of the lungs: update and appraisal, *Toxicol. Appl. Pharmacol.*, 113, pp. 1-12.

Morrow, P.E., Gibb, F.R. and Gazioglu, K.M. (1967), A study of particulate clearance from the human lungs, *Am. Rev. Resp. Dis.*, 96, pp. 1209-1221.

Mossberg, B. (1980), Human tracheobronchial clearance by mucociliary transport and cough, *Europ. J. Respir. Dis.*, Suppl. 107, pp. 51-58.

Nagaishi, C. (1972), *Functional Anatomy and Histology of the Lung*, University Park Press, Baltimore, MD.

National Council on Radiation Protection and Measurements, (NCRP) (1997), Committee on Respiratory Tract Dosimetry Modeling, *Deposition, Retention and Dosimetry of Inhaled Radioactive Substances*, NCRP Report No. 125, Bethesda, MD.

Netter, F.H. (1979), *The CIBA Collection of Medical Illustrations, Volume 7, Respiratory System*, CIBA Pharmaceutical Corporation, Summit, NJ.

Parent, R.A. (Ed.) (1991), *Comparative Biology of the Normal Lung*, CRC Press, Boca Raton, FL.

Patrick, G. and Stirling, C. (1977), The retention of particles in large airways of the respiratory tract, *Proc. Roy. Soc. Lond. B.*, 198, pp. 455-462.

Pavia, D., Bateman, J.R.M. and Clarke, S.W. (1980), Deposition and clearance of inhaled particles, *Bull. Europ. Physiopath. Resp.*, 16, pp. 335-366.

Phalen, R.F., Oldham, M.J., Beaucage, C.B. et al. (1985), Postnatal enlargement of human tracheobronchial airways and implications for particle deposition, *Anat. Rec.*, 212, pp. 368-380.

Saibene, F., Mognoni, P., Lafortuna, C.L. *et al.* (1978), Oronasal breathing during exercise, *Pflugers Arch.*, 378, pp. 65-69.

Schlesinger, R.B. and Lippmann, M. (1976), Particle deposition in the trachea: in vivo and in hollow casts, *Thorax*, 31, pp. 678-684.

Schreider, J.P. and Raabe, O.G. (1981), Structure of the human respiratory acinus, *Am. J. Anat.*, 162, pp. 221-232.

Stanescu, D.C., Pattijn, J., Clement, J. *et al.* (1972), Glottis opening and airway resistance, *J. Appl. Physiol.*, 32, pp. 460-466.

Stöber, W., McClellan, R.O. and Morrow, P.E. (1991), Approaches to modeling disposition of inhaled particles and fibers in the lung, In: *Toxicology of the Lung 2nd Edition* (D.E. Gardner, J.D. Crapo and R.O. McClellan, Eds.), Chapter 19, pp. 527-601, Raven Press, New York.

Stossel, T.P. (1976), The mechanism of phagocytosis, *J. Reticuloendothel. Soc.*, 19, pp. 237-245.

Van As, A. and Webster, I. (1972), The organization of ciliary activity and mucus transport in pulmonary airways, *S. Afr. Med. J.*, 46, pp. 347-350.

Vincent, J.H. (1995), Aerosol science for industrial hygienists, pp. 185–203, Pergamon Press (Elsevier Science Ltd.), Oxford, U.K.

Weibel, E.R. (1963), *Morphometry of the Human Lung*, Academic Press, New York.

Weibel, E.R. and Gil, J. (1977), Structure function relationships at the alveolar level, In: *Bioengineering Aspects of the Lung* (J.B. West, Ed.), Chapter 1. pp. 1-81, Marcel Dekker Inc., New York.

Wood, R.W. (1978), Stimulus properties of inhaled substances, *Environ. Health Perspect.*, 26, pp. 69-76.

Yeh, H.C. and Schum, G.M. (1980), Models of human lung airways and their application to inhaled particle deposition, *Bull. Math. Biol.*, 42, pp. 461-480.

Chapter 3

SAMPLING CRITERIA FOR THE INHALABLE FRACTION

James H. Vincent

Department of Environmental Health Sciences, School of Public Health, University of Michigan, Ann Arbor

3.1 CONCEPT OF INHALABILITY

The first part of the overall process of aerosol exposure is the entry by inhalation (i.e., the aspiration) of aerosols from the ambient air and into the respiratory tract. This part of the exposure process has received relatively little consideration over the years. In fact, it was only as recently as the late 1970s when the idea was first proposed that the only particles which could pose a potential risk to health by inhalation are those which are actually capable of entering the body during breathing (i.e., are inhaled). That is, not all particles will necessarily have the same probability of entering. Thus emerged the concept of the human head as an aerosol sampler and, hence, of *inhalability*. Such thinking has led to a fresh approach to the establishment of relevant guidelines for the sampling of the coarser aerosol fraction.

3.2 EARLIER EXPERIMENTAL MEASUREMENTS OF INHALABILITY

Once it was decided that, in principle, inhalability is an important part of the exposure process, experiments were conducted to obtain quantitative data for the aspiration efficiency of the human head over representative ranges of breathing, aerosol and external wind conditions. Although the original driving force was the practical need to derive criteria for sampling health-related fractions in workplaces, it has always been recognized that the case of the aerosol exposure of humans in the ambient atmosphere is also important. Here an extended range of environmental conditions applies. In particular, whereas the windspeed in workplaces is usually less than 1 ms^{-1} and rarely exceeds 4 ms^{-1}, it could reach 10 ms^{-1} — or even higher — at ground level in the outdoors environment.

The first experimental studies were performed in Britain and Germany in the late 1970s and early 1980s, and were continued into the 1990s. Most were carried out using life-sized human models (in the form of tailors' mannequins) in wind tunnels, generating comprehensive sets of data for the aspiration efficiency of the human head. Such experiments are distinct from those concerned with determining the probability of particle deposition in different parts of the respiratory tract (discussed in other chapters) since they do not require the use of actual human subjects. Since the aerosol and fluid mechanical phenomena which govern inhalability are all external to the human body, these can reliably be simulated by inert experimental systems. In the experiments in question, the mannequins were fitted with breathing machines such that air was inspired cyclically at rates representative of people at work through the nose and/or mouth, and test aerosol was introduced into the airstream upwind of the mannequin. By measuring the concentration of particles upstream of, and of those inhaled by, the mannequin, the aspiration efficiency (A) of the mannequin was measured directly. Here, the quantity A is a fundamental property describing the entry of particles into aerosol sampling devices, and it is defined for each particle size, windspeed, and inhalation flowrate conditions. This quantity was first identified as being applicable to the efficiency by which particles are inhaled by humans by Ogden and Birkett (1977) and so recognized as an important index of human exposure. Thus emerged the concept of the human head as an aerosol sampler. In turn, A is what is now referred to as "*inhalability*."

The experiments reported all employed the same basic methodology for obtaining A, with the test aerosol generated in the wind tunnel upstream of the life-sized mannequin and the upstream reference aerosol concentration obtained from samples collected using isokinetic thin-walled sampling probes. This was compared with the concentration of aerosol as sampled by the mannequin itself (where the 'inhaled' aerosol was collected onto filters mounted just behind the nose and/or mouth).

The earliest such experiments were reported by Ogden and his colleagues (Ogden and Birkett, 1977; Ogden *et al.*, 1977), followed by Vincent and Mark (1982) and Armbruster and Breuer (1982). These covered particle aerodynamic diameters(d_{ae}) up to 100 μm, windspeeds (U) in the range from close to zero to as high as 8 ms^{-1}, and a range of breathing conditions. In the earliest experiments of Ogden and Birkett, A was measured for a range of windspeeds and steady, uniform inspiration flowrates and for a range of mannequin orientations with respect to the wind. The results were used to construct values of A as a function of d_{ae} for a range of cyclical breathing flowrates corresponding to people at work and averaged uniformly over all orientations from 0° to 360°. Armbruster and Breuer also measured A for different windspeeds, cyclical flowrates and orientations. Vincent and Mark measured A specifically for orientation-averaged conditions and for 'at work' breathing conditions only (20 L of air inhaled per minute in 20 breaths), where the mannequin was rotated incrementally during each experimental run. The results of these early studies, for windspeeds ranging from close to zero (calm air) to as high as 8 ms^{-1}, for breathing conditions ranging from those corresponding to 'at rest' to those for 'at work', and for both nose and mouth breathing, are summarized in Figure 3.1. In this figure, the hatched regions shown enclose all the data points obtained in each study.

The results from these three early sets of experiments were notable for the consistency they showed in terms of the broad tendencies exhibited by the data. The main trend for windspeeds below 4 ms^{-1} and for 'at work' breathing conditions was found to be the one between aspiration efficiency (A) and d_{ae}. Detailed inspection of the data reveals that there are some other internal trends with windspeed and breathing conditions, but that these are relatively weak. As highlighted in Figure 3.1, such trends are clearly stronger for windspeeds higher than about 4 ms^{-1}, especially for breathing conditions corresponding to 'at rest.' The overall broad consistency between data sets, especially for conditions thought to be representative of workplaces and workers, was the first hint to suggest the feasibility of applying the overall data towards new health-related definitions of the inhalable fraction of total airborne particulate. Indeed, these earlier data were applied towards the first quantitative inhalability criteria which emerged during the early 1980s (see below) (Vincent and Armbruster, 1981; International Standards Organisation [ISO], 1983; Phalen, 1985).

Most of the major trends found in the earlier studies were repeated in the results of the later work reported by Vincent *et al.* (1990). These too were for orientation-averaged aspiration efficiency, for breathing conditions corresponding to 'at work', and for windspeeds as high as 9 ms^{-1}. These more recent results are shown in Figure 3.2. Here again

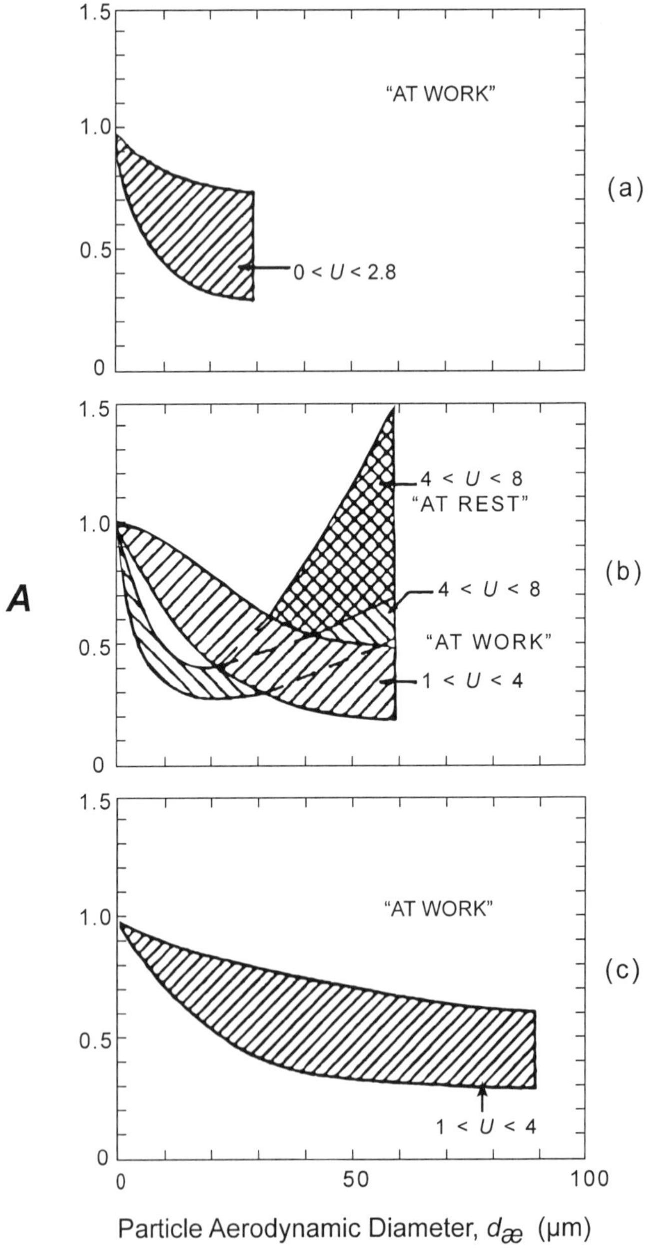

FIGURE 3.1. Summaries of the earlier experimental results for the aspiration efficiency of the human head, based on mannequin studies in wind tunnels (Ogden and Birkett, 1977; Ogden *et al.*, 1977; Vincent and Mark, 1982; Armbruster and Breuer, 1982) (U in ms^{-1}).

the same clear relationship between A and d_{ae} is shown as was observed in the results from the earlier experiments. It is seen that, for the lower windspeeds, A first decreases from unity as d_{ae} increases from zero, then tends to level off at around 0.5 for larger particle sizes. Importantly, there is no evidence of any 'cut-off' in the particle size range indicated. There is, however, a clear effect at the higher windspeeds tested, notably in the appearance of a sharp upturn in A for larger particle sizes.

All the experiments described so far relate to full-grown adults and for a range of levels of physical exertion relevant mostly to people at work. This body of work is supplemented by the study of Phalen *et al.* (1992) which involved wind tunnel experiments to also investigate human aspiration efficiency as a function of body mass, extending down to the range applicable to small children. In this work, Phalen *et al.* invoked the *exposure rate parameter*, the product of inhalability, inhaled minute volume and aerosol mass concentration divided by the body mass, and recommended that this should be applied in the wider context of risk estimation.

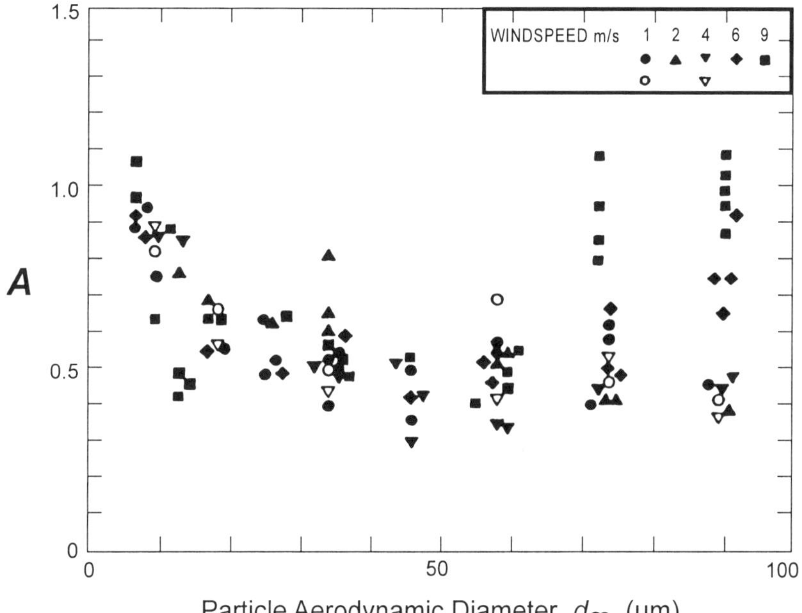

FIGURE 3.2. Later experimental results for the aspiration efficiency of the human head, based on mannequin studies in wind tunnels and a larger range of windspeeds (Vincent *et al.*, 1990).

3.3 More recent experimental measurements of inhalability

The earlier data for the aspiration efficiency — or inhalability — of mannequins in wind tunnels are remarkable for their consistency. Such consistency has provided a major stimulus towards their application in developing a particle size-selective criterion for the inhalable fraction (see below). However, it has always been acknowledged that those experiments did not cover the full range of conditions pertaining to all possible workplaces. So those data were incomplete, most notably at larger particle sizes beyond d_{ae} = 100 μm and low windspeeds below U = 0.5 ms^{-1}. Earlier efforts to widen the range of conditions encompassed were thwarted by the great experimental difficulty involved in suspending very large particles in large wind tunnels at low windspeeds. More recently, however, a number of experimental studies have been reported in which some of those experimental difficulties were challenged and overcome.

Very large particles

Recent studies of particle size distributions in workplaces (Spear *et al.*, 1998) have shown that very large airborne particles may be present, and experience has suggested that such particles may be inhalable. With this in mind, Hinds and his colleagues set out to achieve a spatially uniform aerosol comprising particles with d_{ae} up to 140 μm and to present them to a breathing mannequin in their large wind tunnel. Their results for orientation-averaged aspiration efficiency as a function of d_{ae} for a range of windspeeds from 0.4 to 1.6 ms^{-1} and for breathing conditions corresponding to 'at work' are summarized in Figure 3.3 (Professor W.C. Hinds, communicated privately). These results show that, as in the earlier studies, aspiration efficiency is relatively independent of windspeed. Further, the actual trend in the data is similar to that observed previously. However, the data for d_{ae} in the range from about 30 to 100 μm fall progressively below values characteristic of the earlier data. At d_{ae} = 100 μm, the new measured aspiration efficiency-values lie about 20 percentage points below the earlier ones. As seen in Figure 3.3, the downwards trend continues for larger particles still, although the slope shows a tendency to level off.

Very low windspeeds

Researchers at the United Kingdom Health and Safety Laboratory have reported results in which average windspeeds were investigated at a wide range of industrial workplaces (Berry and Froude, 1989;

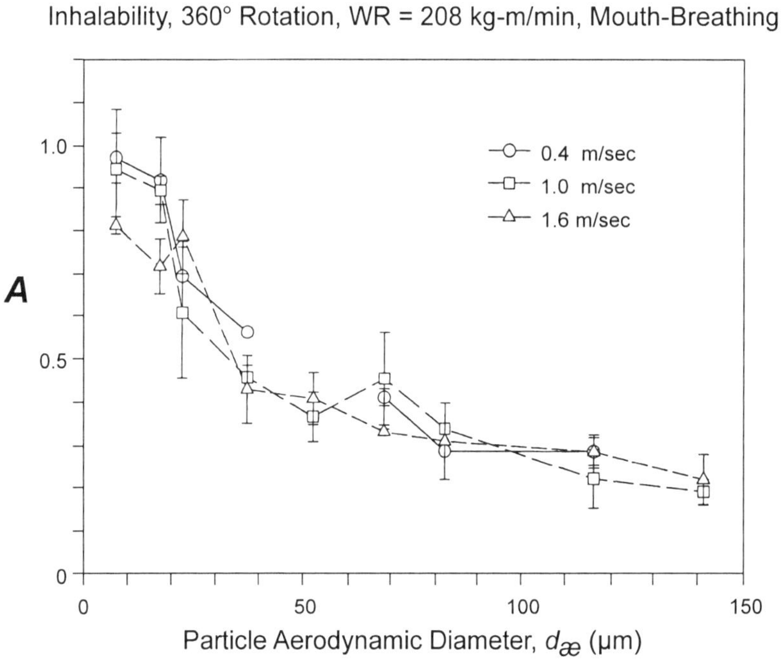

FIGURE 3.3. Recent experimental results (c. 1997) for the orientation-averaged aspiration efficiency of the mouth-breating human head, based on mannequin studies in wind tunnels and a larger range of particle sizes and a single workrate (WR) (Professor W.C. Hinds, personal communication).

Baldwin and Maynard, 1997). These showed that many of the workplaces were characterized by windspeeds lower than the 0.5 ms^{-1} which represented the lower limit of the earlier mannequin studies. This called into some question the validity of relying solely on the earlier wind tunnel results in order to construct a criterion for inhalable aerosol. With this in mind, a new collaborative study was conducted involving two separate laboratories: Health and Safety Laboratory (HSL) and Institute of Occupational Medicine (IOM). As reported by Aitken *et al.* (1999), they set out to overcome the technical difficulties in conducting appropriate experiments at such low windspeed. This was achieved in each laboratory by working with a mannequin in a calm air chamber where the peak local air velocities were found to be no greater than 0.2 ms^{-1}. For the purpose of measuring aspiration efficiency, the reference concentration in the chamber was measured using 'pseudo-isokinetic' probes mounted on a slowly rotating arm. Aspiration efficiency of the mannequin was measured for a range of

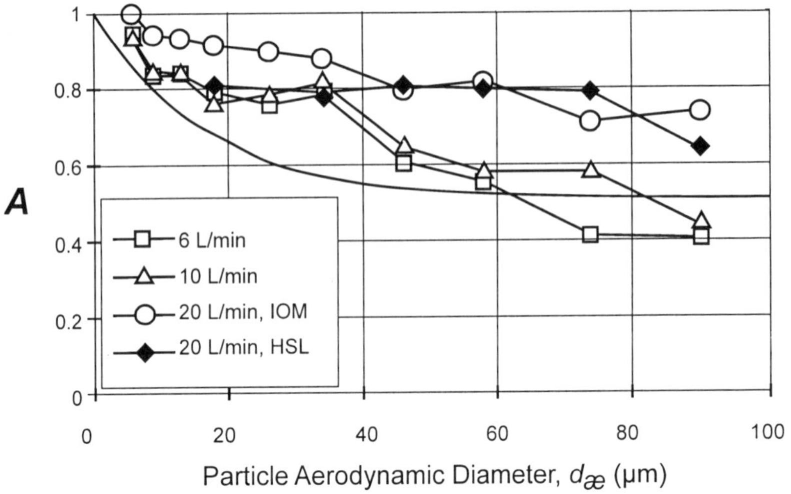

FIGURE 3.4. Recent experimental results for the aspiration efficiency of the human head at very low wind speeds (Aitken *et al.*, 1998).

breathing flowrates from 6 L/min (minute volume, corresponding to 'at rest') to 20 L/min (corresponding to 'at work'); the results as a function of d_{ae} from the two laboratories are summarized in Figure 3.4.

The results in Figure 3.4 are seen to be quite consistent between the two laboratories for conditions which were studied in both. Most striking is the fact that the aspiration efficiency of the mannequin lies markedly above the curve representing the trend exhibited by the earlier data. This tendency is greatest at the highest ('at work') breathing flowrate. This general observed tendency is consistent with the results of the small number of calm-air experiments reported by Ogden and his colleagues as early as 1977. Not shown in Figure 3.4 is the observation that the results reported by Aitken *et al.* showed no significant dependence on whether breathing was through the nose or the mouth. Nor — interestingly — did the results depend on whether or not the mannequin was heated (it having been suggested over the years that the air flow around a living human subject might be influenced by the thermal air currents arising from the elevated temperature of the human body).

Elsewhere, experiments were carried out in the laboratory of the late Professor David L. Swift to examine the aspiration efficiency of nasal-breathing mannequins in calm air (Hsu and Swift, 1999). The mannequins were chosen to simulate both adults and children for breathing corresponding to 'at rest' and 'moderate exercise' respectively. The results are shown in Figure 3.5a and b and exhibit no significant differences between the child and adult mannequins. Both sets of data show that nasal inhalability falls steadily as d_{ae} increases, becoming close to zero for d_{ae} exceeding about 80 μm. Similar experiments were also carried out for actual human subjects, and the results (see Figure 3.5c) are in good agreement with those from the mannequin experiments.

There is a reasonable basis for comparing the results from the Aitken *et al.* and the Hsu and Swift studies. Both were carried out with mannequins in calm air chambers and for similar ranges of particle size and breathing parameters. Both generated data that were very internally consistent and were apparently reproducible. Further, Aitken *et al.* reported that there was no significant difference between nose and mouth breathing. Yet, when the two sets of data from the two separate studies are compared with one another, the difference is striking. The Aitken *et al.* results show that aspiration efficiency is higher than that previously observed for moving air, and the Hsu and Swift results show that aspiration efficiency is lower.

3.4 PHYSICAL BASIS OF INHALABILITY

It is desirable to find a physical explanation for the observed trends in order that the data can be applied with confidence. It is well known from the physical theories for describing the performances of aerosol samplers that aspiration efficiency may be described in general as a function of particle aerodynamic diameter, windspeed, sampling flowrate, sampler body and orifice shapes and sizes and sampler orientation (Vincent, 1989 and 1995). However, at this stage, quantitative predictive modeling of aspiration efficiency has been achieved only for very simple sampler configurations facing the wind. Even today we are not yet close to a rigorous theory that fully accounts for the inhalability observations. Some progress has been made by assuming highly simplistic flow models (including, for example, two-dimensional geometry, perfectly spherical head, unidirectional sampling flowrate, etc.) (Dunnett and Ingham, 1988; Erdal and Esmen, 1995). But, while such mathematical models are instructive and represent progress towards the ideal goal, they are currently of limited practical value. The more realistic human head problem may be simplified somewhat by assuming fixed shape and dimensions (i.e., a 'standard' human head) and by imposing fixed cyclical breathing conditions (i.e., 'at work'). Then, blunt sampler theory based on the semi-empiri-

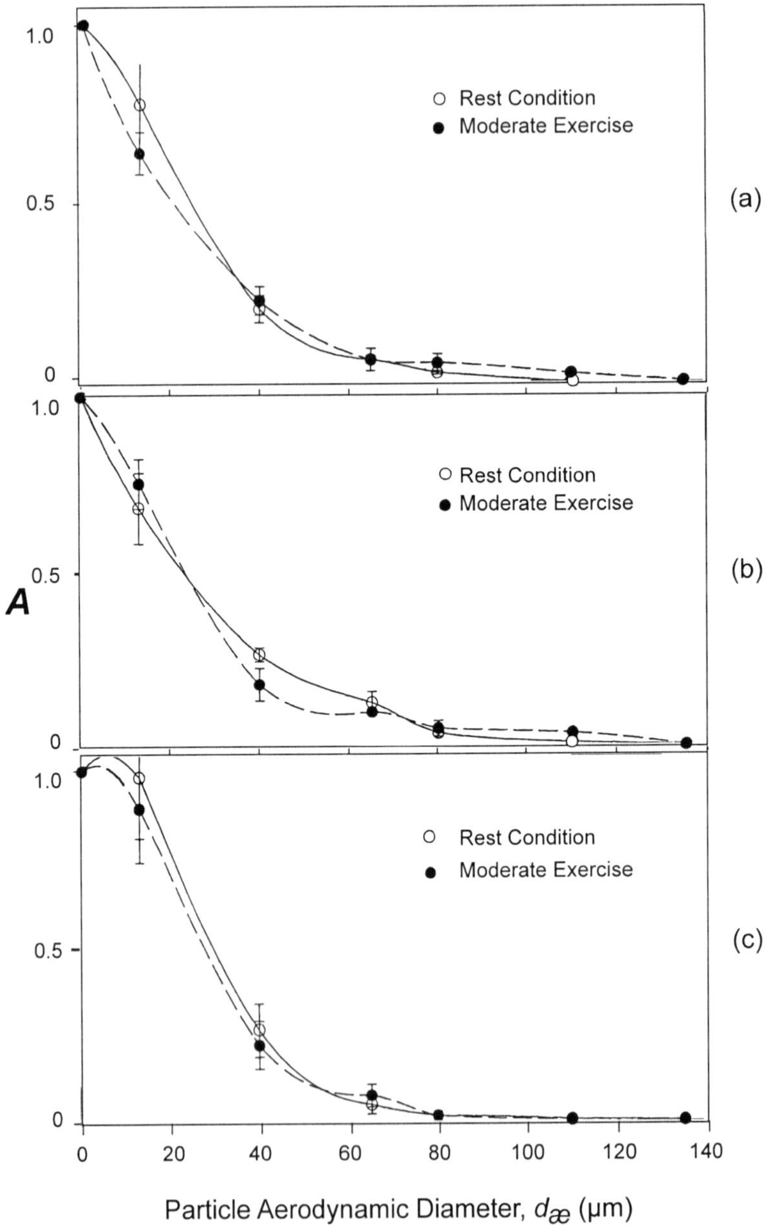

FIGURE 3.5. Recent experimental results for the aspiration efficiency of the human head at very low windspeeds, for adult and child mannequins (a and b) and for actual adult human subjects (c), all for nasal breathing, for 'at rest' and 'moderate exercise' breathing conditions (Hsu and Swift, 1998).

cal so-called 'impaction-model' approach permits the functional statement (Vincent and Mark, 1982; Vincent, 1989; Vincent *et al.*, 1990)

$$A = f\{ K, U \} \qquad \textbf{3.1}$$

where

$$K = d_{ae}^{\;2}\, U \qquad \textbf{3.2}$$

is an inertial parameter which reflects the ability of particles to follow the streamlines of a distorted air flow just outside the nose and mouth. It is, in effect, a stop distance for particles moving near the human head acting as an aerosol sampler or, alternatively, a Stokes' number (*St*) in which the dimensional scale is fixed. To explore how the experimental data relate to Equation 3.1, the aspiration efficiency data from the later experiments are re-plotted in Figure 3.6 in a form that appears to give a fair 'collapse.' In the absence of a quantitative physical model, we have to fall back on an empirical expression for describing the main observed trends. The form

$$A \exp\{\mathrm{a}(1 - U)\} \equiv \mathrm{A}^* = [\exp\{b\, K^c\} + \mathrm{d}\, K] \qquad \textbf{3.3}$$

is consistent with Equation 3.1 but has physical meaning only in that *A* is a function of *K* and *U* as suggested by aspiration theory. When this is applied to the experimental data in Figure 3.6, best-fit values for the coefficients are

$$a = 0.0267,\ b = -0.0393,\ c = 0.370 \text{ and } d = 1.04 \times 10^{-5}$$

so long as d_{ae} is expressed in [μm] and *U* in [m/s].

The observed trends for mouth breathing are plausible within the framework of what is known about the physics of the sampling process. This plausibility includes not only the tendency for there to be no cut-off in *A* for the particle size range indicated but also the tendency for *A* to increase as the windspeed increases and/or the aspiration flowrate decreases. Both derive from the fact that, as windspeed increases, a greater fraction of larger particles impact — by virtue of their inertia — onto the mouth of the mannequin when it is oriented in forward-facing directions. Further insight can be obtained by considering the aspiration efficiency of a thin-walled tube facing into the wind. Here it is noted that the aspiration efficiency at very large particle sizes tends to level off at a value equal to the ratio between the windspeed and the average air velocity over the sampling inlet (Vincent, 1989). It should be noted, however, that the physical scenario in

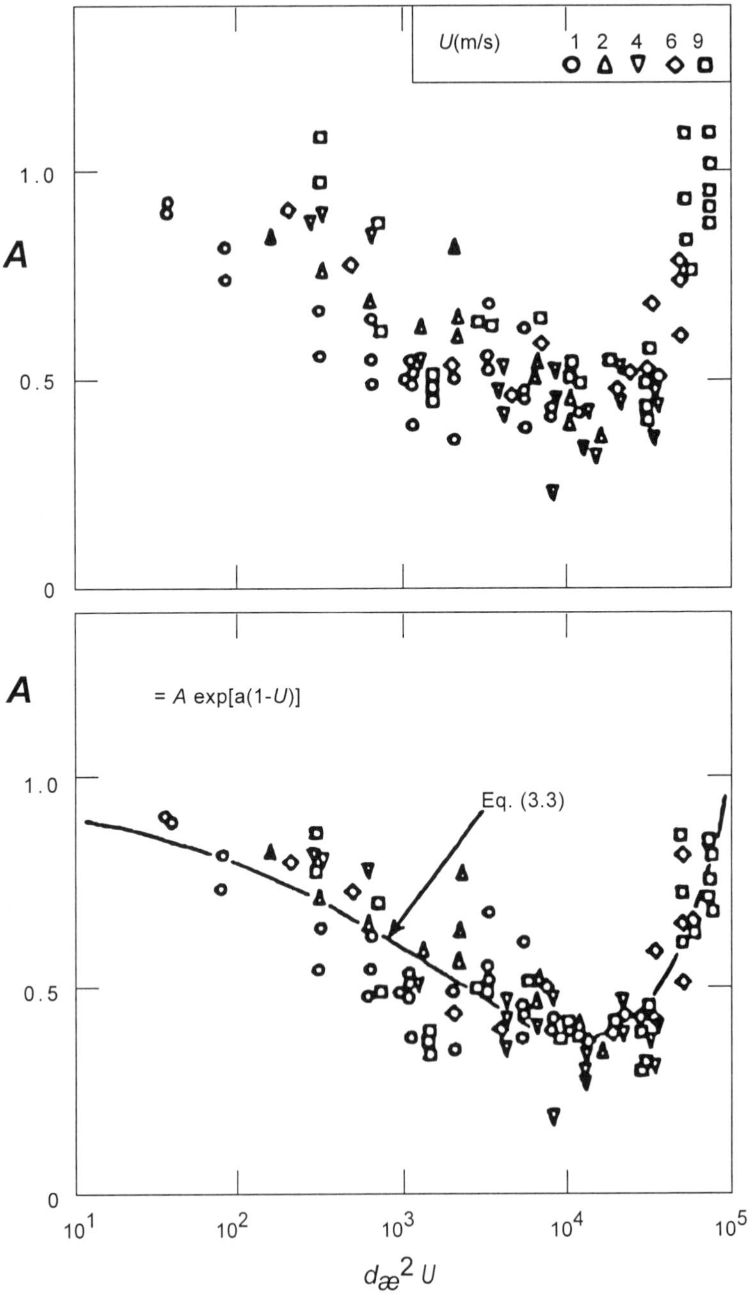

FIGURE 3.6. Plot of the Vincent *et al.* (1990) experimental data with respect to relevant physical parameters, where the observed 'collapse' is indicative of the role of inertia. The solid line on the lower graph is the best fit represented by Equation (3.3).

question assumes that particle inertia is the dominant physical mechanism governing aspiration and that the effect of gravity is negligible. The latter assumption may not be true for very large particles or for very low windspeeds. So the interpretation of aspiration efficiency results like those shown needs to be treated with some caution. It is intuitively clear, for example, that a particle size must eventually be reached where particles are falling so quickly under the influence of gravity that they cannot possibly be inhaled — in which case A must eventually fall to zero. Although that situation was never reached in any of the earlier reported experiments (upon which the preceding explanation is based), it certainly became a feature in some of the more recent experiments, most notably those performed at very low windspeeds. In this regard, the nose-breathing results of Hsu and Swift (1999) are the more consistent in relation to what might be expected from theory.

3.5 Progress towards criteria for the inhalable fraction

For some types of aerosol, particles constitute a risk to health, regardless of where they are eventually deposited. So, in the first place, inhalable aerosol is a fraction which is an objective for measurement in itself in many environmental and occupational situations. In the past, the recommendations for the health-related sampling of coarse particles in most countries have been — and continue to be — based on the concept of so-called 'total' aerosol. This concept is intended to relate to all particulate matter that might be considered airborne, and it contains the underlying assumption that all particles which are airborne have the same probability of being inhaled. As seen from the data presented above, where inhalability varies with d_{ae}, this assumption can no longer be supported. Meanwhile, however, practical sampling instruments for 'total' aerosol have continued to be sold commercially and extensively deployed in industrial hygiene. Most such instruments were originally developed without particular regard to specific quantitative criteria or performance indices, and now — upon closer inspection — it is becoming apparent that their performance characteristics have varied greatly from one to the other. It follows, therefore, that what we refer to as "total aerosol" has been effectively defined in each particular situation by the particular sampling instrument chosen to do the job. In addition, switching from one instrument to another in a given practical situation might well produce different measurements of exposure, even though the level of exposure itself might not have actually changed. With the preced-

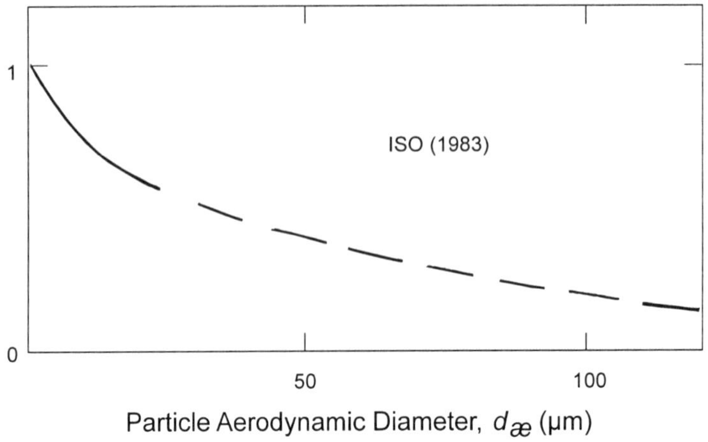

FIGURE 3.7. The original 1983 convention for inhalable aerosol as proposed by the International Standards Organisation (ISO).

ing in mind, the concept of inhalability has emerged as important in unifying aerosol exposure in a scientific manner which is strongly related to how aerosols enter the bodies of people during breathing (and hence to health).*

Data from experiments like those described above have formed the basis of recommendations for replacing the old 'total' aerosol concept with a quantitative sampling convention based on human inhalability. The first — and historically important for that reason — was that proposed by the ISO in 1983. During the meetings of its ad hoc working group in the late 1970s, the only data available for consideration were those from the experiments of Ogden *et al.* (1977) and Ogden and Birkett (1977) for d_{ae} up to about 30 µm. These data were extrapolated to larger particle sizes and used to arrive at the curve described by the purely empirical expression for inhalability, $I(d_{ae})$:

$$I(d_{ae}) = 1 - 0.15\,\{\log_{10}(1 + d_{ae})\}^2 - 0.10\,\log_{10}(1 + d_{ae}) \qquad \textbf{3.4}$$

where d_{ae} is expressed in [µm] and the aspiration efficiency for the human head (A) has been replaced by I to indicate that it has now been applied to a convention for aerosol sampling. This curve is shown in Figure 3.7, where the dashed part of the line indicates where the

*Elsewhere in this book, whenever the term "total aerosol" does not refer to all particles that are in the air, single quotes are used; i.e., the term 'total' is used.

curve has been extrapolated from the actual experimental data. In the 1983 ISO recommendations it was stated that, when this curve may be used as a criterion for practical sampling purposes, a tolerance band may be applied where d_{ae} may be allowed to vary by ±15% at each prescribed value. Further latitude may be permitted if it can be demonstrated that 67% of mass samples obtained in practice fall within ±10% of the result that would be achieved had Equation 3.4 been followed exactly.

Somewhat later, the more comprehensive data set, including in particular results for larger particles with d_{ae} up to about 100 μm obtained from subsequent experimental studies (Vincent and Mark, 1982; Armbruster and Breuer, 1982), provided the basis for the proposal of a more representative curve (Vincent and Armbruster, 1981). The main difference in the new, modified curve from the earlier ISO one is the absence of a cut-off at large particle sizes and — instead — the leveling-off in I at around 0.5. In the 1985 ACGIH recommendations, this curve was included as a convention for defining the inhalable (known as 'inspirable' at the time) fraction and described by the empirical expression

$$\text{IPM}\ (d_{ae}) = 0.5\ \{1 + \exp(-\,0.06\ d_{ae})\} \qquad \textbf{3.5}$$

for d_{ae} (again expressed in [μm]) up to and including 100 μm. Beyond 100 μm, it was acknowledged that there was no information on which to base a firm recommendation. It should be noted, however, that this does not imply a 'cut-off', as the 1985 ACGIH report is sometimes misunderstood as saying. The curve given by Equation 3.5 is shown in Figure 3.8. In its original 1985 convention, the ACGIH also recommended that, as a performance band for practical sampling instruments, I may vary by ±0.10 for each value of d_{ae}. That is, as with the ISO recommendations, there is a working envelope which is representative of a high proportion of the original data and so is a reasonable basis for assessing the performances of practical devices.

In the original ACGIH proposal referred to, attention was focused primarily on particle size-selective criteria for sampling in indoor workplaces where windspeeds even as high as 4 ms^{-1} are uncommon. For most workplace conditions, therefore, curves like those in the two conventions described above may be regarded as generally adequate. However, there are some outdoor workplaces where conditions might sometimes lie outside these wind conditions. Furthermore, at least as far as the ISO is concerned, sampling in the ambient atmosphere for the purpose of evaluating the risk to the community at large is an important part of its general charge. From the more recent experi-

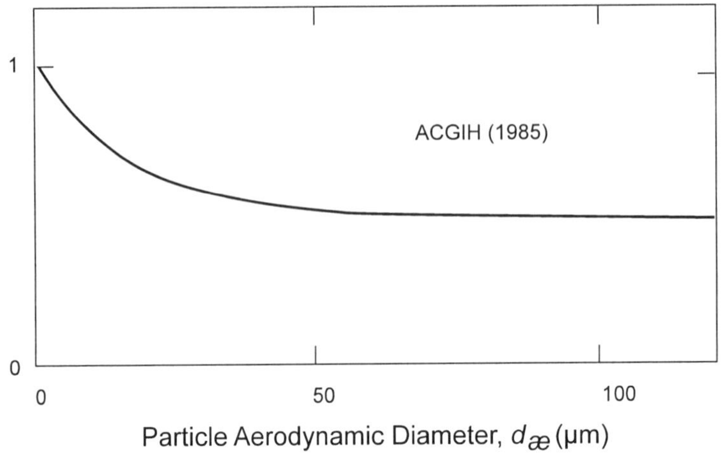

FIGURE 3.8. The 1985 convention for inhalable aerosol as proposed by ACGIH (as currently agreed now — for low windspeeds — by ISO and CEN).

mental evidence shown earlier in Figures 3.1 and 3.2, it is clear that the existing ISO and ACGIH conventions do not properly represent what happens at those higher windspeeds. In particular, the experimental data suggest that using samplers with performance based on either of these curves could lead to a significant underestimation of the exposure of humans to large particles. This could be important in situations where there are large particles containing potentially hazardous substances (e.g., radioactive nuclides, heavy metals, polycyclic aromatic hydrocarbons, etc.). It is therefore appropriate to consider how the definition of inhalability might be extended.

As shown earlier, Equation 3.3 represents the experimental data for the aspiration efficiency of the human head quite well over the range of conditions examined. But it is substantially more complicated than both the existing conventions shown, most notably the 1985 ACGIH convention as shown in Equation 3.5 which is the one most representative of the majority of the data now available. So Equation 3.3 is not really suitable for direct application as a practical criterion for the performance of sampling instruments. In the search for a more suitable generalized expression, the 1985 ACGIH curve provides perhaps the best starting point. With this in mind, a new convention was proposed, based on a simple empirical modification of the 1985 ACGIH curve, of the form (Vincent *et al.*, 1990)

$$\text{IPM}\ (d_{ae}) = 0.5\ \ \{1 + \exp(-0.06\ d_{ae})\} + B(d_{ae}, U) \qquad \textbf{3.6}$$

where the new, additional term (B) is given by

$$B(d_{ae}, U) = p\, U^{q} \exp\{ r\, d_{ae}\} \qquad 3.7$$

in which d_{ae} is again in [μm] and the windspeed U is in [ms^{-1}]. The coefficients p, q, and r are constant, for which best fit with the experimental data are achieved for

$$p = 1 \times 10^{-5}\ ,\ q = 2.75, \text{ and } r = 0.055.$$

The new convention is shown graphically in Figure 3.9 where the experimental data of Vincent *et al.* (1990) are also plotted. The curves are seen at low windspeed ($U < 4\ ms^{-1}$) to be almost identical to the ACGIH curve, and so it is applicable to most workplace situations. At higher windspeed, however, it also gives good agreement with the trends exhibited by the latest experimental data obtained under those conditions. In general, therefore, the new expression as described by Equation 3.7 corresponds well with the experimental data over the full range of conditions. Recently this generalized expression has been incorporated into a revised ISO set of proposals (ISO, 1992). The version applying to lower windspeeds, where Equation 3.7 reverts back to Equation 3.5, is also embodied in the criteria adopted by the Comité Européen de Normalisation (CEN, 1992).

3.6 CONCLUSIONS AND RECOMMENDATIONS

From the extensive research carried out using life-sized mannequins, a large data base has been developed for the aspiration efficiency of the human head, or inhalability (I), for ranges of windspeeds and breathing flowrates relevant to people at work. These results, from more than one research group in more than one laboratory in Europe and North America, are sufficiently consistent to allow generalization of the results in the form of a simplified convention for inhalability (I) as a function of particle aerodynamic diameter (d_{ae}). Recognizing that inhalability under certain external wind conditions is a function of windspeed, it is recommended that, for the purpose of a sampling convention, inhalable particulate matter (or IPM, replacing what was referred to above as "I") should be expressed in the general empirical form

$$\mathrm{IPM}(d_{ae}) = 0.5\ \{1 + \exp(-0.06\ d_{ae})\} + 10^{-5} U^{2.75} \exp(0.055) \qquad 3.8$$

in which d_{ae} is given in [μm] and the windspeed U is given in [ms^{-1}]. This convention reverts back to the simpler form for IPM given, reiterating Equation 3.5, by

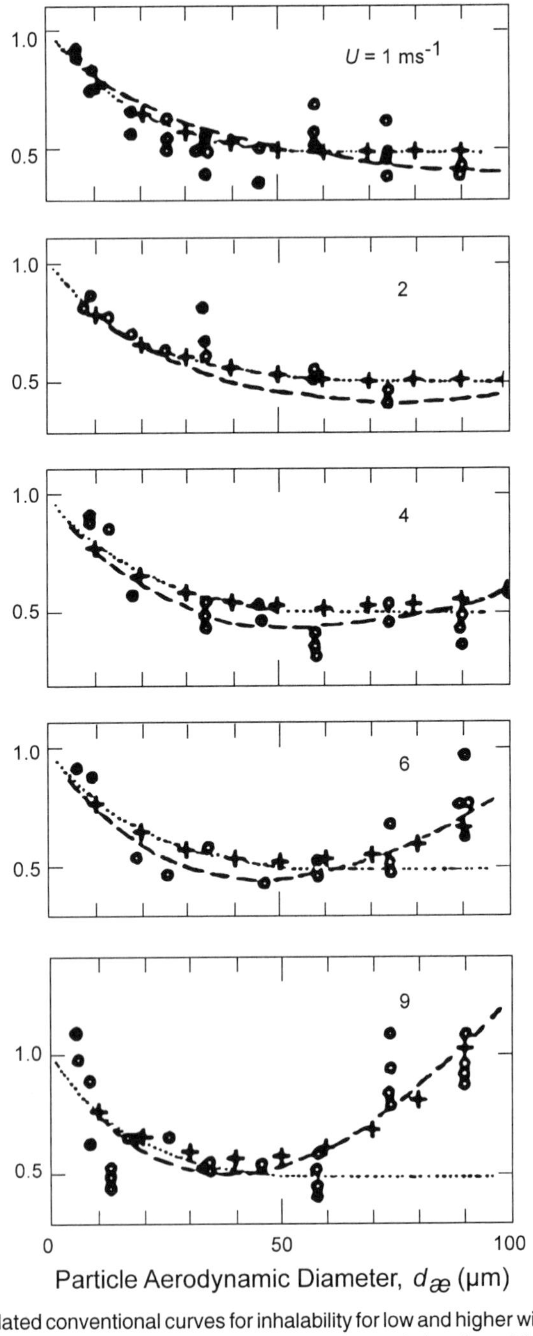

FIGURE 3.9. Calculated conventional curves for inhalability for low and higher windspeeds, obtained using Equation (3.8); also shown are the experimental data from Vincent *et al.* (1990).

$$\mathrm{IPM}(d_{ae}) = 0.5 \bullet \{1 + \exp(-0.06\ d_{ae})\} \qquad \mathbf{3.9}$$

when the windspeed U falls below about 4 ms^{-1}. It is important to note that, while the above applies only for d_{ae} up to 100 µm, it does not imply the existence (or otherwise) of a 'cut-off' at d_{ae} = 100 µm. Numerical values for inhalability arising from Equations 3.8 and 3.9 are listed in Table 3.1.

TABLE 3.1. Table of numerical values for inhalability, as a function of particle aerodynamic diameter (d_{ae}) and for a range of windspeeds (U), as calculated from Equation 3.8.

d_{ae} (µm)	U = 0.5 ms^{-1}	U = 1 ms^{-1}	U = 4 ms^{-1}	U = 9 ms^{-1}
0	1.00	1.00	1.00	1.00
2	0.94	0.94	0.94	0.95
4	0.89	0.89	0.89	0.90
6	0.85	0.85	0.85	0.85
8	0.81	0.81	0.81	0.82
10	0.77	0.77	0.78	0.78
12	0.74	0.74	0.74	0.75
14	0.72	0.72	0.72	0.72
16	0.69	0.69	0.69	0.70
18	0.67	0.67	0.67	0.68
20	0.65	0.65	0.65	0.66
25	0.61	0.61	0.61	0.63
30	0.58	0.58	0.59	0.60
35	0.56	0.56	0.56	0.59
40	0.55	0.55	0.55	0.58
45	0.53	0.53	0.54	0.58
50	0.52	0.53	0.53	0.59
55	0.52	0.52	0.53	0.61
60	0.51	0.51	0.53	0.63
65	0.51	0.51	0.53	0.66
70	0.51	0.51	0.53	0.71
75	0.51	0.51	0.53	0.77
80	0.50	0.50	0.54	0.85
85	0.50	0.50	0.55	0.95
90	0.50	0.50	0.57	1.10
95	0.50	0.50	0.59	1.28
100	0.50	0.50	0.61	1.53

Regarding tolerance bands for the proposed inhalability criterion, it is recommended that, for the purpose of the convention and application of the inhalability (or IPM) criterion as a standard against which to compare a sampling instrument, the following requirements should be met:

(a) The majority of the performance data — in terms of sampling efficiency for a given particle size and windspeed — for the instrument in question should fall within ± 0.10 of the inhalability values indicated for each particle size; and

(b) The total sampled mass, as measured using the instrument in question, should — under the conditions of practical use — fall within ± 20% of that which would be measured using a hypothetical ideal instrument whose performance perfectly matches the inhalability curve.

It is important to reiterate that the preceding empirical expressions are applicable only over the range of the conditions for which the experimental data were obtained; namely $0 < d_{ae} < 100$ µm and $0.5 < U < 9$ ms^{-1}. Until recently, no information was available outside those ranges. But lower windspeeds in particular, and perhaps also larger particles, may be relevant to many workplaces. As reported above, new experiments are starting to shed some light about the trends that might be expected. Aitken *et al.* (1998) did in fact go so far as to tentatively suggest a new inhalability criterion for very low windspeed having the empirical form

$$IPM\ [d_{ae}] = 1 - 0.0038\ d_{ae} \qquad \textbf{3.10}$$

where, as before, d_{ae} is in [µm]. However, in view of the disparity between the Aitken *et al.* and the Hsu and Swift experimental results, this suggestion should be regarded with caution until further experimental data are available. Although, therefore, it is at present too early to make firm recommendations to take such conditions into account, the possibility that a revised criterion might be proposed at some time in the future is not ruled out.

REFERENCES

Aitken, R.J., Baldwin, P.E.J., Beaumont, G.C. et al. (1999), Aerosol inhalability in very low winds, *J. Aerosol Sci.*, in press.

Armbruster, L. and Breuer, H. (1982), Investigations into defining inhalable dust, In: *Inhaled Particles V* (W.H. Walton, Ed.), Pergamon Press, Oxford, U.K., pp. 21-32.

Baldwin, P.E.J. and Maynard, A.D. (1997), Measurement of windspeeds in indoor workplaces, Proceedings of the 11th Annual Aerosol Society Conference, The Aerosol Society, Bristol, U.K.

Berry, R.D. and Froude, S. (1989), An investigation of wind conditions in the workplace to asses their effect on the quantity of dust inhaled, U.K. Health and Safety Executive Report IR/L/DS/89/3, Health and Safety Executive, London, U.K.

Comité Européen de Normalisation (CEN) (1992), Workplace atmospheres: size fraction definitions for measurement of airborne particles in the workplace, CEN Standard EN 481.

Dunnett, S.J. and Ingham, D.B. (1988), An empirical model for the aspiration efficiencies of blunt aerosol samplers oriented at an angle to the oncoming flow, *Aerosol Sci. Tech.*, 8, pp. 245-264.

Erdal, S. and Esmen, N.A. (1995), Human head model as an aerosol sampler: calculations of aspiration efficiencies for coarse particles using an idealised human head model, *J. Aerosol Sci.*, 26, pp. 253-272.

Hsu, D.-J. and Swift, D.L. (1999), The in vitro measurements of human inhalability of ultra-large aerosols in calm air conditions, *J. Aerosol Sci.*, in press.

International Standards Organisation (ISO), *Air quality — particle size fraction definitions for health-related sampling*, Technical Report ISO/TR/7708-1983 (E), ISO, Geneva, 1983, revised 1992.

Ogden, T.L. and Birkett, J.L. (1977), The human head as a dust sampler. In: *Inhaled Particles IV* (W.H. Walton, Ed.), Pergamon Press, Oxford, U.K., pp. 93-105.

Ogden, T.L., Birkett, J.L. and Gibson, H. (1977), *Improvements to dust measuring techniques*, IOM Report No. TM/77/11, Institute of Occupational Medicine, Edinburgh, Scotland, U.K.

Phalen, R.F. (Ed.) (1985), *Particle size-selective sampling in the workplace*, Report of the ACGIH Air Sampling Procedures Committee, American Conference of Governmental Industrial Hygienists (ACGIH), Cincinnati, OH.

Phalen, R.F., Oldham, M.J. and Dunn-Rankin, D. (1992), Inhaled particle mass per unit body mass per unit time, *Appl. Occup. Environ. Hyg.*, 7, pp. 246-252.

Spear, T.M., Werner, W.A., Bootland, J. *et al.* (1998), Assessment of particle size distributions of health-related aerosol exposures of primary lead smelter workers, *Ann. Occup. Hyg.*, 42, pp. 73-80.

Vincent, J.H. (1989), *Aerosol Sampling: Science and Practice*, Wiley and Sons, Chichester, U.K.

Vincent, J.H. (1995), *Aerosol Science for Industrial Hygienists*, Pergamon Press, Elsevier Science, Oxford, U.K.

Vincent, J.H and Armbruster, L. (1981), On the quantitative definition of the inhalability of airborne dust, *Ann. Occup. Hyg.*, 24, pp. 245-248.

Vincent, J.H. and Mark, D. (1982), Application of blunt sampler theory to the definition and measurement of inhalable dust, In: *Inhaled Particles V* (W.H. Walton, Ed.), Pergamon Press, Oxford, U.K., pp. 3-19.

Vincent, J.H., Mark, D., Miller, B.G. *et al.* (1990), Aerosol inhalability at higher windspeeds. *J. Aerosol Sci.*, 21, pp. 577-586.

◆ ◆ ◆ ◆ ◆

Chapter 4

SAMPLING CRITERIA FOR THE THORACIC AND RESPIRABLE FRACTIONS

Otto G. Raabe[1] and Bruce O. Stuart[2]

[1]*University of California at Davis, Institute of Toxicology and Environmental Health, Davis, CA*

[2]*Adjunct Professor of the Graduate School, Department of Pharmaceutical Sciences, University of Connecticut, Storrs, CT*

4.1 Introduction

The behavior of inhaled airborne particles in the respiratory airways and their alternative fates of either deposition in the various airway regions or exhalation depend upon the physical behavior of the aerosol particles under the specific physiological and anatomical conditions (Hatch and Gross, 1964; Raabe, 1982). The dynamics of the behavior of particles that are carried with the flow of air into the respiratory airways depend primarily upon the particle size distribution and the aerodynamic properties of the particles. Other characteristics may alter the properties of the particles or influence their behavior, including their electrostatic charge, hygroscopicity or deliquescence, and chemical composition. The geometry of the airways from nose or mouth to the lung parenchyma also influences aerosol particle deposition. Important morphometric parameters describing the respiratory airways include the diameters, lengths, and branching angles of airway segments. Physiological factors include breathing patterns, air flow dynamics in the airways, and variations in relative humidity and temperature within the respiratory tract. Utilizing the specific details of these controlling factors, theoretical models of regional deposition can be developed to predict the fate of inhaled particles of various types. Carefully collected data from experiments with human volunteers provide the basis for testing such predictions.

The aerodynamic properties of aerosol particles depend upon several physical characteristics, including size, shape, and physical densities of the particles. Two important properties of aerosol particles are their inertia, which applies most appropriately to particles with diameter larger than about 0.5 μm, and their diffusional properties which become increasingly important for particles decreasing the range of diameter below about 0.5 μm in diameter (related to the particle diffusivity) (Raabe, 1976 and 1980). The inertial properties are describable in terms of an equivalent particle aerodynamic diameter, for which particles with identical inertial properties have the same aerodynamic diameter irrespective of their actual sizes, shapes or densities. This equivalent diameter (dae) is defined (Hatch and Gross, 1964) as the diameter of a sphere of density 103 kg/m3 (the same as that of pure water) having the same settling speed under gravity in still air as the particle in question, whatever its shape and density (Raabe, 1976).

Bombardment of airborne particles by gas molecules in air produces random (Brownian) motion which causes very small particles (especially those smaller than 0.5 μm) to move and mix even under externally tranquil conditions. This property of small particles is described in terms of a *diffusive diameter*, *ddf*, defined (Raabe, 1980) as equal to the geometric diameter of an ideal spherical particle with the same diffusivity as the actual particle under identical conditions. Regional deposition of inhaled particles in the respiratory tract depends primarily on the particle aerodynamic diameter for larger particles and primarily on the diffusive diameter for smaller particles.

Even with carefully developed theoretical models, and with the support of reliable deposition data, it is not possible to predict exactly the quantitative regional deposition of particles of a given inhaled aerosol in a particular person. Biological variability between individuals, differences in health, confounding factors such as cigarette smoking, and differences that relate to age or breathing styles, as well as inherent differences in airway sizes, can cause differences in the fraction of an inhaled aerosol that may deposit in the airways. In turn, such factors influence the quantity of particulate mass that is actually inhaled by a given individual. For the purposes of occupational health protection, reasonable predictions must be made using the available models and data employing certain simplifying assumptions concerning biological factors. A degree of conservatism is desirable in estimating the expected deposition of inhaled particles or penetration of particles into the airways. That is, it is best to use models that would tend on the average to somewhat overestimate the potential inhalation risks. However, realistic estimates are needed to prevent expensive over-regulation that is not really justified. These considerations need to be carefully weighed in establishing appropriate size-selective sampling criteria. For the general predictions of respiratory tract deposition of inhaled airborne particles used in this text, the biological parameters describing the "*reference worker*" (see Chapter 2) are used to provide the basis for describing representative biological conditions for a person under moderately active work conditions.

Very early in the study of airborne dusts in occupational settings such as coal mining, it became clear that the incidence of specific respiratory disease associated with occupational exposure to dust does not always correlate well with the total airborne particulate mass concentration. This is because particles of different sizes deposit with different patterns and probabilities in the various airway regions. Large particles tend to deposit primarily in the head airways region (HAR) or tracheobronchial region (TBR). Therefore, large particles tend to have less effect than smaller particles on the small airways of the deep lung. Thus, total aerosol alone does not adequately predict these differing effects. Respirable dust standards were developed specifically to separate the bigger particles from the smaller particles in a manner similar to the size-selective particle collection of the upper airways. So samples of dusts (such as those containing silica) taken in this way are indicative of the particles actually available to the deep lung during inhalation (Lippmann, 1970a). These size-selective samples do not actually indicate the deposition of respirable particles in the lung (see Chapter 11), but rather they provide a measure of the particulate mass available to the deep lung during breathing. Because the larger particles are unlikely to reach the deep lung and contribute to small airway disease, they are sampled in a manner that makes them similarly unlikely to be collected. Hence, particle size-selective sampling involves separating the sample aerodynamically into two portions using: 1) a simple pre-collection device to remove larger particles according to prescribed criteria; followed by 2) a tandem filter to collect the small particle sample for assay. No sharp 'cut-size' is intended since no specific size is distinctively collected by respiratory tract regions. Instead, particle size-dependent criteria are defined for selective sampling in order to divide an aerosol sample into two fractions that simulate the size fractionation which normally occurs in the respiratory airways themselves. Maximum permissible concentrations of dust that are based upon this appropriately biased size-selective sample are clearly more relevant in inhalation hazard evaluation than gross samples of total airborne particulate mass.

In some cases, particles of a particular type may be of biological concern even if they deposit in the large airways of the lung, while in other cases important risks may even be associated with deposition of relatively large particles in the nose or other HAR. If only one type of size-selective sample is collected, there may be difficulties in interpreting risks that relate to the TBR versus the HAR. The *thoracic particulate matter* (TPM) criterion is designed to complement the *respirable particulate matter* (RPM) criterion and provide information relevant to particles that may have adverse effects in large as well as small airways of the lung. The *inhalable particulate matter* (IPM) criterion (see Chapter 3) provides the basis for evaluating the concentrations of airborne particles that may have adverse effects anywhere in the body including the HAR.

This chapter reviews selected deposition models and experimental data and provides recommendations for the sampling characteristics of size-selective samplers for collecting TPM and RPM in working and living environments. TPM refers to the portion of the inhaled airborne particles that may penetrate the head airways and enter the lung airways under the worst-case condition of inhaling through the mouth. RPM refers to that portion of the inhaled airborne particles that may penetrate the head and tracheobronchial airways during nasal breathing and enter the gas exchange region (GER) of the lung.

4.2 Respiratory Tract Deposition of Inhaled Particles

Inhaled particles that pass through the HAR of the respiratory tract during inhalation may deposit in the TBR or the GER of the lung. The magnitude of this deposition will depend upon aerodynamic and diffusive particle sizes, respiratory mechanics, and anatomical relationships. The deposition of particles associated with inhalation via the mouth rather than the nose is particularly important because the protective deposition of larger particles in the nasopharyngeal (NP) region is much higher for all sizes than the deposition that occurs in the oropharyngeal head airways during mouth breathing. Specifically, much larger particles that normally are collected in the NP portion of the head airways during breathing via the nose may pass the glottis and larynx and deposit in the lung during breathing via the mouth.

Inertial impaction is the dominant mechanism of deposition of particles with d_{ae} larger than 3 µm in the HAR or TBR. In this process the airborne particles, because of their inertia, do not follow changes in direction or speed of air streamlines and they may collide with and attach to the wall of the airway. For example, if air is directed towards an airway surface (such as a branch carina) but the forward velocity is suddenly reduced because of the change in flow direction caused by the obstruction of the surface, inertial momentum may carry larger particles across the air streamlines and onto the moist surface of the tract where they are deposited. Aerodynamic separation of this type may be characterized in terms of an impaction parameter given approximately by d_{ae}^2Q where Q is the average inspiratory flow rate.

Deposition by diffusion can also occur in the nose, mouth and pharyngeal airways for very small particles (smaller than 0.01 µm) as a result of the Brownian motion of these ultra-fine particles. This effect is not of great significance for most aerosols encountered in the environment and in occupational situations, so it is customary to neglect diffusional deposition of aerosols in the upper airways. Diffusional deposition in the GER of the lung is, however, the predominant collection mechanism for particles smaller than 0.5 µm in physical diameter.

In order to size-selectively sample aerosols for estimating the fraction or mass concentration of particles available for deposition in conducting airways or GER of the lung during inhalation, it is necessary to estimate the fraction of inhaled aerosol that can pass the nose and nasal pharynx or mouth and oral pharynx as a function of particle size. These estimates can be made using mathematical models of expected deposition with respect to particle size of particles in the NP and oralpharyngeal portions of the HAR or from experimental data obtained for human subjects inhaling well characterized particles under controlled conditions.

Until recent years, the most widely used models of regional deposition versus particle size were developed by the International Commission on Radiological Protection (ICRP) Task Group on Lung Dynamics under the chairmanship of Professor P.E. Morrow (Morrow *et al.*, 1966). Additional data from more recent studies have been incorporated into updated deposition models published by the ICRP in 1994 and the National Council on Radiation Protection and Measurements (NCRP) in 1997. Although the purpose of these models was to determine radiation exposure from inhaled radioactive aerosols, the ICRP and NCRP aerosol deposition models are broadly applicable to environmental and occupational aerosols. Morrow *et al.* (1966)

used representative values for normal respiratory parameters, although it is understood that these values may not describe any particular person. Considerable variability in respiratory parameters may occur among individuals in the population, particularly when healthy adults are contrasted with children. The ICRP reference values for a typical adult worker are given by Snyder (1975), including body weight of 70 kg, height of 175 cm and body surface area of 1.8 m^2. The ICRP Task Group used the anatomical model and general methods of Findeisen (1935) and Landahl (1950) for calculating deposition in the tracheobronchial and pulmonary regions. Particles were assumed to be insoluble, stable, and spherical with physical densities of 10^3 kg/m^3 (the same as for pure water). Regional deposition was calculated for a presumed average-sized person for a breathing rate of 15 breaths per minute via the nose for three tidal volumes: a) 0.75 L, corresponding to 'at rest'; b) 1.45 L, corresponding to 'moderate activity'; and c) 2.15 L, corresponding to fairly strenuous 'at work' activity. Their result for 1.45 L is summarized in Figure 4.1.

Morrow *et al.*(1966) calculated the theoretically expected deposition of particles in the airways of the lung during breathing, but they did not theoretically determine head airway deposition. Their NP deposition calculations were based upon the empirical loglinear equation of Pattle (1961)

$$\text{Deposition fraction} = -0.62 + 0.475 \log_{10}[d_{ae}^2 Q] \qquad \textbf{4.1}$$

for NP deposition during inspiration where $\log_{10}$ is the ten base logarithm. They assumed this equation also applied to the exhalation process to calculate the total NP deposition. By assuming that particles up to d_{ae} = 10 μm reach the TBR region during mouth breathing, the ICRP Task Group results can be used to estimate deposition of particles in the GER during mouth breathing or TBR deposition for particles entering the trachea.

Particle deposition data in humans have been collected from volunteers inhaling test aerosols through either mouthpieces or nose tubes utilizing monodisperse, insoluble, stable aerosols of different sizes. The most extensive of these studies are those of Lippmann and Albert (1969), Giacomelli-Maltoni *et al.* (1972) and Heyder *et al.* (1975). Additional useful data have been reported by Altshuler *et al.* (1967), George and Breslin (1967), Hounam *et al.* (1971), Lippmann (1970b and 1977), Shanty (1974), Foord *et al.* (1976), Heyder and Rudolf (1977), Chan and Lippmann (1980), and Stahlhofen *et al.* (1980). Results for total respiratory deposition are compared to the ICRP Task Group models in Figures 4.2 and 4.3, respectively, for nose and mouth

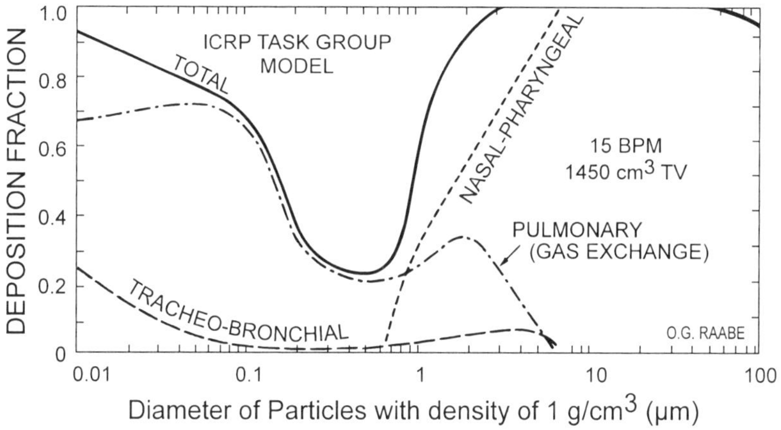

FIGURE 4.1. Total and regional deposition fractions for various sizes of inhaled airborne spherical particles with physical density of 1.0 g/cm^3 in the human respiratory tract as calculated by the International Commission on Radiological Protection (ICRP) Task Group on Lung Dynamics (Morrow *et al.*, 1966) for nasal breathing at a rate of 15 breaths per minute (BPM) and tidal volume (TV) of 1450 cm^3 (Raabe, 1982).

breathing. Deposition in the TBR for particles entering the trachea is summarized in Figure 4.4 and deposition in the GER during mouth breathing in Figure 4.5.

The major differences in deposition that occur between nose breathing (Figure 4.2) and mouth breathing (Figure 4.3) are demonstrated by the theoretical and experimental information of deposition in the lung GER during oral breathing (Figure 4.5). When aerosols are inhaled through the nose, the relatively efficient filtration action of the NP region (see Figure 4.6) prevents the passage of particles with d_{ae} larger than 10 μm to the lung. This markedly limits the pulmonary (gas exchange) region deposition of particles between 2 μm and 10 μm in aerodynamic diameter. An active person (reference worker) breathing at 15 breaths per minute with a tidal volume of 1.45 L (see Figure 4.1) would be expected to deposit in the deep lung (GER) about 35%, 25%, 10%, and essentially 0% of inhaled particles with d_{ae} equal to 0.2 μm, 1 μm, 5 μm and 10 μm respectively, during nasal breathing. Likewise, the tracheobronchial deposition would be expected to be about 2%, 3%, 6% and 0% respectively for these same particle sizes.

Mouth breathing markedly alters the deposition of inhaled particles in humans in that larger particles can enter both the TBR (see Figure 4.4) and the GER (see Figure 4.5). The deposition in the deep lung (GER) would be expected to increase to about 35%, 30%, 55% and

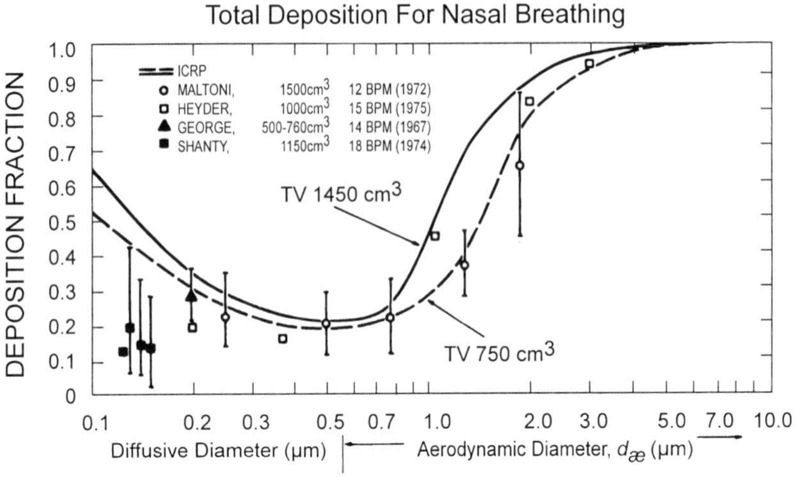

FIGURE 4.2. Selected data reported for the deposition in the entire respiratory tract of monodisperse aerosols inhaled by persons via the nose from Giacomelli-Maltoni *et al.*, (1972); Heyder *et al.*, (1975); George and Breslin, (1967); and Shanty (1974); compared with predicted values calculated by the ICRP Task Group on Lung Dynamics (Morrow *et al.*, 1966) for tidal volumes (TV) of 750 cm^3 (dashed line) and 1450 cm^3 (solid lines) (Raabe, 1982).

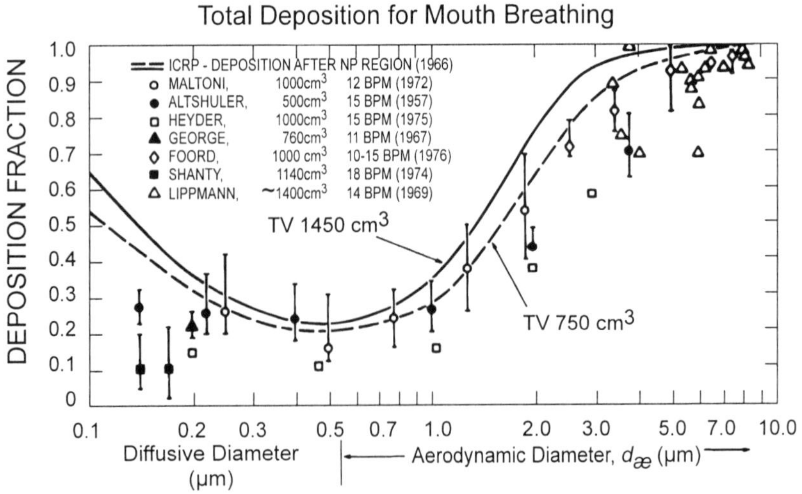

FIGURE 4.3. Selected data reported for the deposition in the entire respiratory tract of monodisperse aerosols inhaled by persons through the mouth from Giacomelli-Maltoni *et al.*, (1972), Altshler *et al.*, (1967), Heyder *et al.*, (1975), George and Breslin (1967), Foord *et al.*, (1976), Shanty, (1974), and Lippmann and Albert (1969), compared with predicted values calculated by the ICRP Task Group on Lung Dynamics (Morrow *et al.*, 1966) for tidal volumes (TV) of 750 cm^3 (dashed line) and 1450 cm^3 (solid lines) (Raabe, 1982).

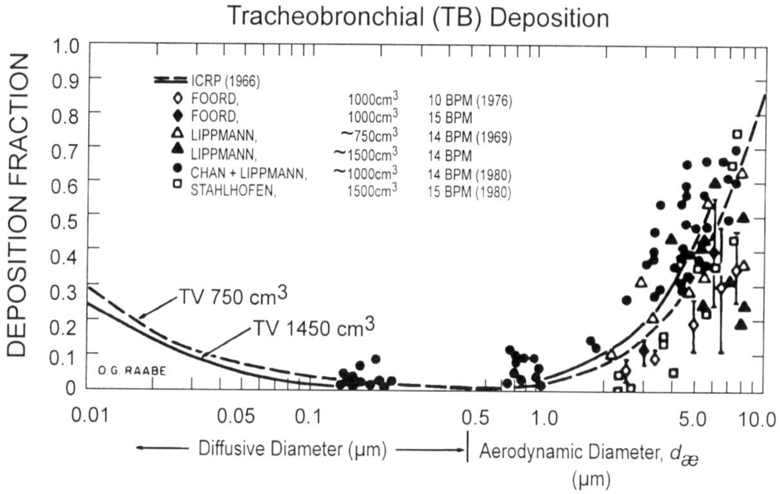

FIGURE 4.4. Selected data reported for the deposition in the human tracheobronchial region (TBR) as fraction of monodisperse aerosols entering the trachea reported by Foord *et al.*, (1976), Lippmann and Albert (1969), Chan and Lippmann (1980), and Stahlhofen *et al.*, (1980) compared with predicted values calculated by the ICRP Task Group on Lung Dynamics (Morrow *et al.*, 1966) for tidal volumes (TV) of 750 cm^3 (dashed line) and 1450 cm^3 (solid lines) (Raabe, 1982).

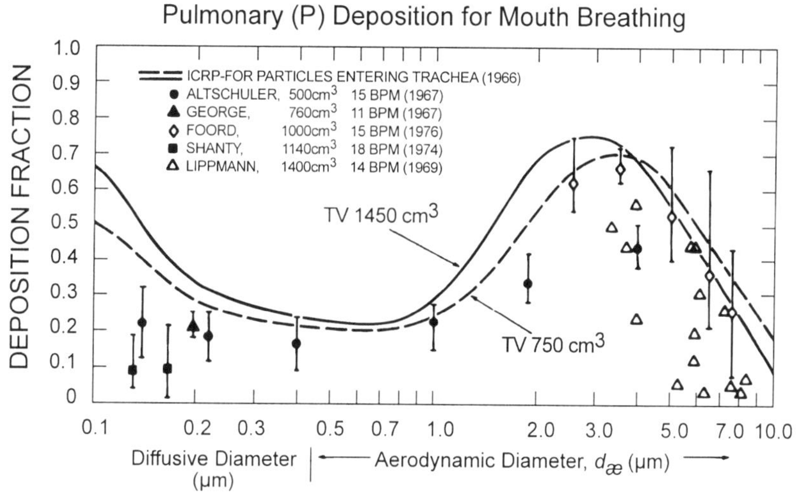

FIGURE 4.5. Selected data for pulmonary (P) gas exchange region (GER) deposition of monodisperse aerosol inhaled through the mouth by people reported by Altshuler *et al.*, (1967), George and Breslin (1967), Foord *et al.*, (1976), Shanty (1974), and Lippmann and Albert (1969) compared with predicted values calculated by the ICRP Task Group on Lung Dynamics (Morrow *et al.*, 1966) for tidal volumes (TV) of 750 cm^3 (dashed line) and 1450 cm^3 (solid lines) (Raabe, 1982).

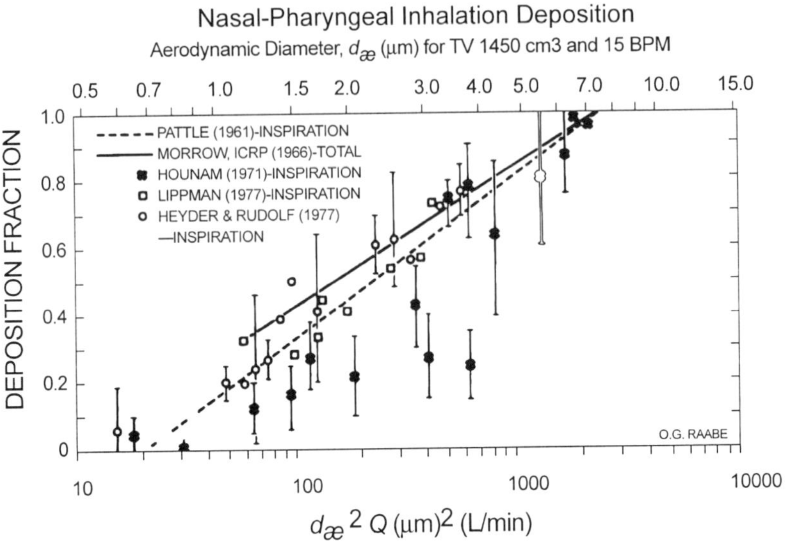

FIGURE 4.6. Selected data reported for inspiratory deposition in the nasal-pharyngeal (NP) portion of the head airways region (HAR) of monodisperse aerosols inhaled by people through the nose as reported by Hounam *et al.*, (1971), Lippmann (1977), and Heyder and Rudolf (1977), compared to the empirical model of Pattle (1961) for inspiration alone (dashed line) as a function of the impaction parameter $d_{ae}^2 Q$ where d_{ae} is the aerodynamic diameter and Q is the average inspiratory flow rate. Also shown are the corresponding aerodynamic diameters for reference worker for tidal volume of 1450 cm^3 and 15 breaths per minute (BPM) and the calculated total deposition fraction for both inspiration and exhalation given by the ICRP Task Group on Lung Dynamics (Morrow *et al.*, 1966).

10% respectively for inhaled particles with d_{ae} equal to 0.2 µm, 1 µm, 5 µm and 10 µm for the reference worker breathing via the mouth at 15 breaths per minute with a tidal volume of 1.45 L.

Relating the preceding to what is discussed in Chapter 3, for many workplace environments the inhalable particulate mass concentration may not be significantly larger than the thoracic particulate mass concentration due to the removal of the largest particles from workplace air by normal sedimentation and impaction before inhalation occurs. Thoracic deposition in the tracheobronchial airways (TBR) and deep lung (GER) is emphasized during mouth breathing. This suggests that the thoracic sample should be based upon the particles available to the whole lung (TBR plus GER) during mouth breathing. An estimate of the fraction of particles available to the lung during mouth breathing requires an estimate of the deposition of particles in the mouth, oral pharynx, and other parts of the head airways during

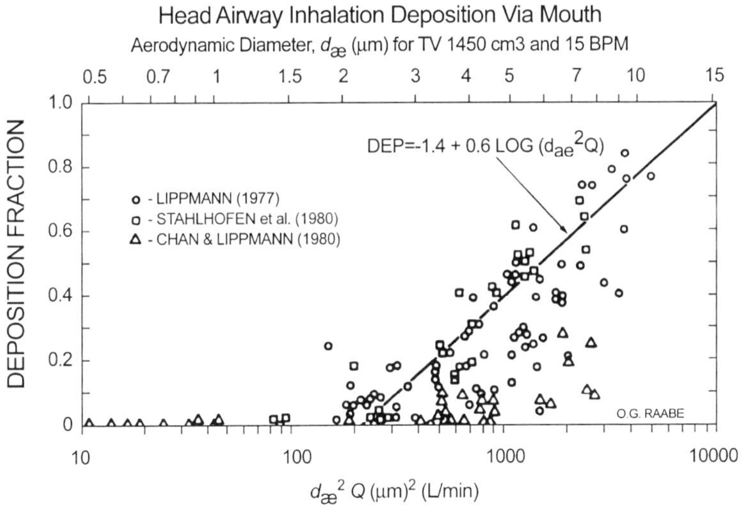

FIGURE 4.7. Selected data reported for inspiratory deposition in the head airways region (HAR) of monodisperse aerosols inhaled by people through the mouth as reported by Lippmann (1977), Stahlhofen *et al.*, (1980), and Chan and Lippmann (1980) as a function of the impaction parameter d_{ae}^2Q where d_{ae} is the aerodynamic diameter and Q is the average inspiratory flow rate. Also shown are the corresponding aerodynamic diameters for reference worker for tidal volume (TV) of 1450 cm^3 and breathing rate of 15 breaths per minute (BPM) equivalent to Q = 43.5 L/min.

mouth breathing. The ICRP Task Group provided no predictions of head deposition during mouth breathing, but experimental data are available from the experiments of Lippmann and Albert (1969), Lippmann (1977), Chan and Lippmann (1980), and Stahlhofen *et al.* (1980). These are summarized in Figure 4.7 along with the empirical fitted function

$$\text{Deposition fraction} = -1.4 + 0.6 \log_{10}[d_{ae}^{\ 2}Q] \qquad \textbf{4.2}$$

which approximately fits selected portions of the data.

4.3 Thoracic particulate matter (TPM)

Thoracic particulate matter (TPM) represents those airborne particles that, by virtue of their aerodynamic size and airborne properties, are expected to penetrate the head airways beyond the larynx and enter the lung airways (TBR and GER) during mouth breathing. TPM therefore represents the highest potential exposure of the whole lung to inhaled particles.

The injurious changes in the TBR caused by deposited inhaled particulate materials produced in certain workplace environments may include bronchoconstriction, edema of bronchial walls, and cancer. The localized patterns of asbestos fiber deposition in bronchiolar airway bifurcations appear to be associated with the initial sites of bronchial cancer (Lippmann, 1990; McClellan *et al.*, 1992). Concomitant cigarette smoking may greatly increase the risk of such bronchogenic carcinoma. Other probable primary causative agents that have been implicated in this fatal disease are inhaled compounds of nickel, chromium, arsenic, beryllium, and coal tar (coke oven and roofing tar environments). Among uranium and other underground miners, the radioactive polonium atoms (^{218}P0 and ^{214}Po) produced by the decay of the radon gas are attached to inhaled carrier particles and are deposited in the upper tracheobronchial airways. Lining epithelial cells in these airways at the fourth to sixth generations receive the highest radiological doses and appear to be the site of origin of bronchogenic carcinoma (Harley *et al.*, 1984).

The TPM fraction of any aerosol may be obtained by passing the airborne particles through an inertial separator designed to collect the larger particles but passing the smaller particles in a manner that has size penetration efficiencies based on the penetration of the head airways that occurs during mouth breathing. The particles that penetrate the inertial separator are collected on a filter, or other media, and gravimetrically or chemically assayed to determine the mass of material or quantity of specific chemical agent associated with the thoracic sample. These quantities are converted to a concentration value for the sampled atmosphere by division by the total volume of air sampled. The aerosol penetrating the thoracic separator therefore represents the fraction of airborne particles that is available to the whole lung because of penetration of the head airways during mouth breathing. Depending upon the size distribution of the particles in the sample, a portion is deposited in the TBR of the lung. A further portion of the particles in the thoracic fraction is deposited in the GER of the lung, and a portion is exhaled without being deposited.

The sampling criterion for the thoracic fraction size-selective sampler must consider the shape of the actual airway penetration curve as a function of particle size and the types of samplers available to simulate that penetration curve. Cyclone separators are frequently used since they tend to have performance characteristics with shapes qualitatively similar to airway penetration curves. Several groups have suggested a 'cut point' at $d_{ae} = 10$ µm, representing the 50% efficiency point for such a thoracic sampler, including the U.S. Environmental Protection Agency (*Federal Register*, March 9, 1984) and the International Standards Organisation (ISO, 1981).

It is convenient to describe the performance of such a separator (as well as the shape of the actual airway particle collection data) as a cumulative lognormal function with distribution of efficiencies from near zero to approximately 100% described in terms of a geometric standard deviation about the median 'cut size.' Based upon review of the data and the foregoing considerations, the TPM criterion for particles entering the mouth during oral inhalations is established as a curve consisting of those particles, already pre-selected as meeting the IPM criterion (as defined in Chapter 3), that penetrate a separator whose size collection efficiency is described by a cumulative lognormal function with median ($_{50}d_{ae}$) of 11.64 µm aerodynamic diameter and with geometric standard deviation of $\sigma_g = 1.5$.

This thoracic curve is shown in Figure 4.8 as penetration efficiency versus particle aerodynamic diameter. Also shown in Figure 4.8 are selected data from Figure 4.7 but corrected for the reference worker (by adjusting the inertial parameter for flow rate $Q = 43.5$ L/min) for head penetration of particles inhaled by people via the mouth in controlled experiments described by Lippmann (1977), Stahlhofen *et al.* (1980), and Chan and Lippmann (1980). The recommended curve lies to the right of the data corrected to the $Q = 43.5$ L/min of the 'reference worker,' suggesting that the actual penetration to the lung tends to be less in most cases than the TPM particle size-selective sample. The TPM criterion therefore tends to somewhat overestimate the amount of exposure of the lung and correspondingly to provide a degree of protection when used as the basis of risk estimates. The overall cut point of a sampler that corrects for both TPM and IPM falls at the $d_{ae} = 10$ µm using these criteria.

4.4 Respirable particulate matter (RPM)

Injurious changes following the deposition of inhaled particles in the pulmonary (gas exchange) region may be caused by cadmium, osmium tetroxide, bauxite fumes, beryllium and a variety of organic materials. Cadmium oxide fumes that produce interstitial pneumonitis can also produce emphysema and other chronic pulmonary diseases. Several insoluble dusts are known to cause pulmonary fibrosis following prolonged exposure, including aluminum and asbestos. Silicosis is caused by the chronic inhalation and deposition of crystalline silica in this region (Stuart *et al.*, 1985). Asbestosis results from the prolonged daily exposure to and deposition of asbestos fibers in the GER, causing interstitial fibrosis, calcification and fibrosis of the pleura (Lippmann, 1990). The progressive massive fibrosis associated with "black lung" disease is caused by chronic deposition of respirable

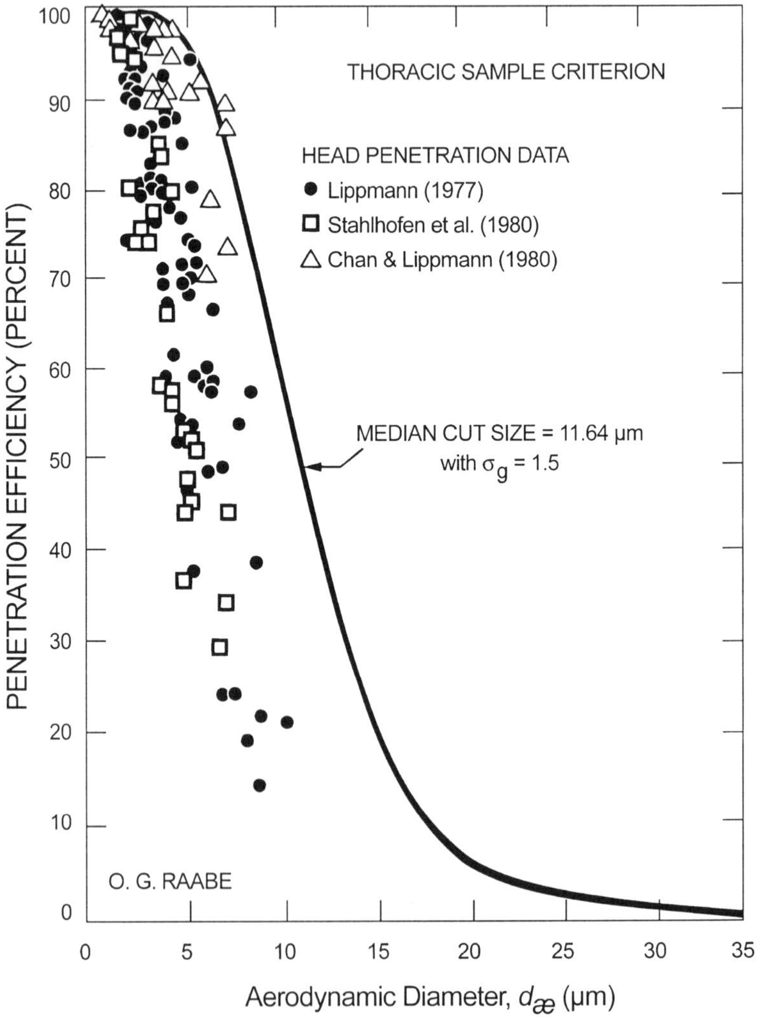

FIGURE 4.8. Thoracic particulate matter (TPM) sample criterion for particles entering the mouth during oral inhalation given as sample collection efficiency for those particles that penetrate a separator whose size collection efficiency is described by a cumulative lognormal function with median cut size of 11.64 µm aerodynamic diameter and with geometric standard deviation of 1.5. Also shown are selected data for observed human head penetration during inhalation by mouth corrected to the appropriate aerodynamic diameter for reference worker (Q = 43.5 L/min).

coal mine dust particles in the lungs of coal miners under conditions of inadequate ventilation.

For occupational situations where the risk of inhalation exposure is primarily associated with deposition of insoluble particles in the deep lung, ACGIH previously (since 1968) recommended a respirable dust standard. In this context, the word "*respirable*" has been used to describe that portion of an aerosol available to the GER during nose breathing. The earlier ACGIH respirable sample standard was based upon those particles that penetrate a sampler whose penetration curve has a value (fraction) of 90% for particles smaller than d_{ae} = 2 µm, 50% at 3.5 µm, 25% at 5 µm, and 0% at 10 µm. Similar respirable dust standards have been proposed and used by the U.S. Atomic Energy Commission (AEC) and the British Medical Research Council (BMRC) (Lippmann, 1970a) as shown in Figure 4.9. Both cyclone separators and horizontal elutriators have been used to provide the appropriate penetration characteristics. The objective is to make use of a sampler to collect particles that would be deposited in the NP section and TBR during nose breathing in approximately the same proportion as their actual deposition with respect to particle size. The particles that penetrate the inertial separator are collected on a filter, or other media, to provide the size fraction representing the particles available to the GER of the lung.

Using the latest ACGIH respirable dust curve as the basis, the RPM criterion is established as a curve describing those particles, already preselected as meeting the IPM criterion, that penetrate a separator whose size collection efficiency is described by a cumulative lognormal function with median ($_{50}d_{ae}$) of 4.25 µm with geometric standard deviation (σ_g) of 1.5. This curve is shown in Figure 4.10 along with a curve representing NP and TB penetration. The RPM size-selective sampling criterion incorporates and clarifies the respirable dust standard. Although there are no direct measurements of the penetration to the GER of particles inhaled via the nose, the recommendations of the Task Group on Lung Dynamics (ICRP, 1966) for NP deposition (Figure 4.6) and for tracheobronchial deposition of particles entering the trachea (Figure 4.4) are in reasonable agreement with the data collected by Lippmann and Albert (1969), Foord *et al.*(1976), Lippmann (1970b and 1977), Hounam (1971), Heyder and Rudolf (1977), Chan and Lippmann (1980), and Stahlhofen *et al.*(1980). The sum of the Task Group deposition predictions for the nasalpharyngeal region (NPR) and TBR were used to calculate an effective penetration curve which is shown in Figure 4.10. It can be seen that the RPM criterion somewhat overestimates the fraction of particles available to

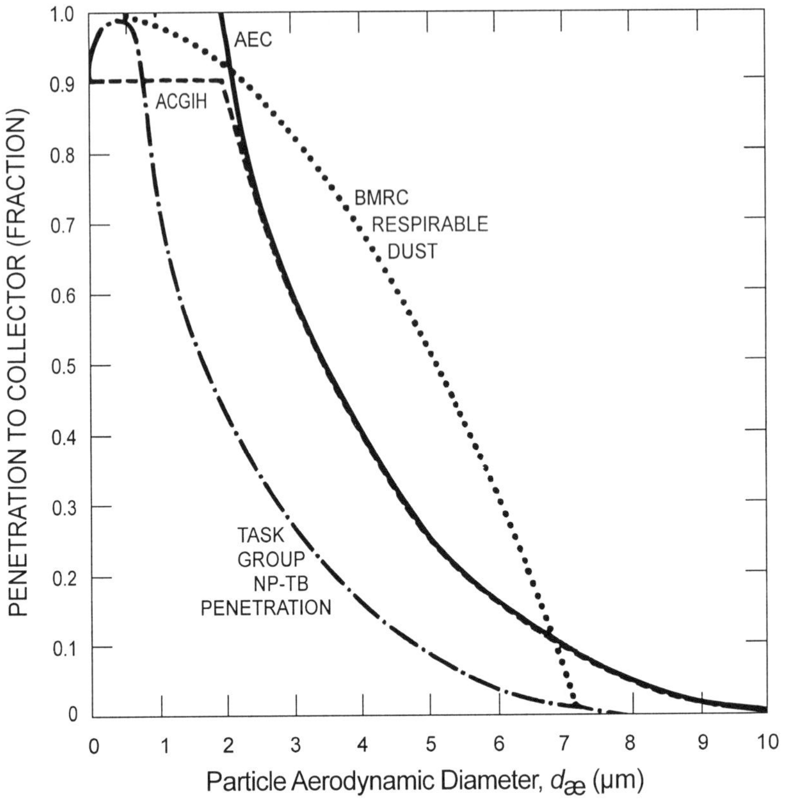

FIGURE 4.9. Respirable dust sample criteria given as fractional penetration to collector versus aerodynamic diameter based upon the British Medical Research Council (BMRC), the U.S. Atomic Energy Commission (AEC) recommendations, and the original American Conference of Governmental Industrial Hygienists (ACGIH) recommendations (Lippmann, 1970a), compared with estimated penetration values for nasalpharyngeal (NP) and tracheobronchial (TB) airways based on Morrow et al. (1966).

the deep lung (GER) when compared to the Task Group-derived penetration curve and therefore provides a degree of protection in assessing inhalation hazards. The overall cut point of a sampler that corrects for both RPM and IPM falls at $d_{ae} = 4$ µm using these criteria.

4.5 Discussion

Although both the TPM and RPM curves are to the right of the average respiratory airway penetration values for head airway penetration and combined nasalpharyngeal – tracheobronchial penetration,

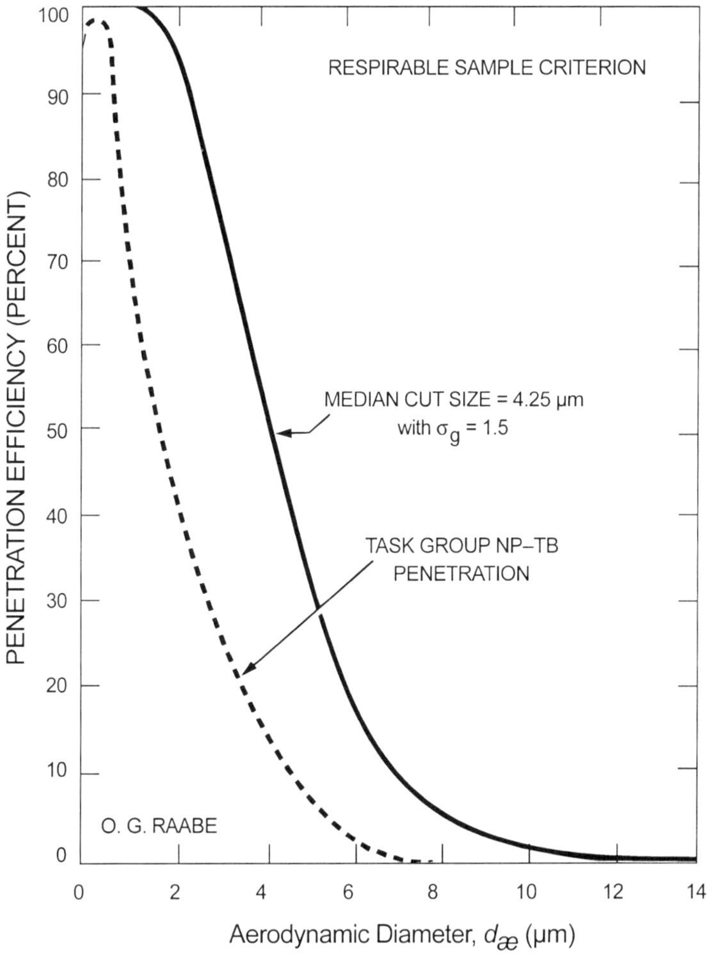

FIGURE 4.10. Respirable particulate matter (RPM) sample criterion for particles entering the nose during nasal inhalation given as sample collection efficiency for those particles that penetrate a separator whose size collection efficiency is described by a cumulative lognormal function with median cut-size of 4.25 µm aerodynamic diameter and with geometric standard deviation of 1.5. Also shown are estimated penetration values for nasal pharyngeal (NP) and tracheobronchial (TB) airways based on Morrow et al. (1966).

respectively, this does not really represent an unwarranted overprotection. This is because many individuals would be expected to have actual upper airway penetrations that exceed the average values and may actually exceed the curves in some isolated cases. Also the deposition data were evaluated from observed impaction data in terms of the parameter d_{ae}^2Q and converted to the specified aerodynamic diameters by assuming $Q = 43.5$ L/min for the 'reference worker.' In many cases, the actual breathing volumetric flow rate may be much lower than this value, reaching only one-half this value for some individuals in some situations. The particle sizes that would have been calculated would have been larger for lower values of Q, so the deposition data would have been to the right of the positions shown in Figures 4.8 and 4.10. On the other hand, oral breathing is associated primarily with higher breathing rates so that the use of the active reference worker flow rate Q of 43.5 L/min is most appropriate with the head airway deposition for oral breathing or oronasal breathing.

Inherent in the descriptions of thoracic and respirable sampling given above is the assumption of 100% inhalability for these samples. The IPM criterion (see Chapter 3) suggests that this assumption is not strictly correct. On the average, the human head is not an ideal sampler, so that less than 100% of the total airborne particulate matter is actually available to the lung airways (TBR plus GER). In the case of the RPM fraction that consists of particles generally smaller than $d_{ae} =$ 4 μm in aerodynamic diameter, RPM sampling as described would tend to slightly overestimate the respirable mass concentration. Similarly, because the TPM fraction consists primarily of particles smaller than $d_{ae} = 10$ μm in aerodynamic diameter, thoracic size-selective sampling as described will tend to overestimate the thoracic particulate mass concentration, typically by up to an average of about 15%. Implicit in the criteria for both TPM and RPM is the fact that an inhalable pre-selector is located upstream of the TPM or RPM selector itself. In general, it is recommended in practice that no special correction is needed for either of these size-selective samples. However, when a mass balance is being evaluated among the IPM, TPM and RPM, possible small discrepancies may need to be considered. Indeed, they may become significant for coarse aerosols. There may also be small differences between samples collected with slightly different instruments. Instruments cannot be designed to meet exact criteria, so small differences among different types of instruments will occur.

In summary, airborne IPM can be measured as a function of aerodynamic equivalent diameter, d_{ae}, using a sampler that collects particles with a size-selective fractional efficiency, $SI(d_{ae})$, given by:

$$SI\,(d_{ae}) = 0.5 + 0.5 \exp\,[-0.06\, d_{ae}] \qquad \textbf{4.3}$$

Selected values of the IPM fraction are given as a percentage in Table 4.1.

TABLE 4.1. Selected values of inhalable, thoracic and respirable particulate matter fractions (in %) as functions of particle aerodynamic equivalent diameter, d_{ae}.

Particle Aerodynamic Diameter (μm)	Inhalable (IPM) Particulate (%)	Thoracic (TPM) Particulate (%)	Respirable (RPM) Particulate (%)
0	100	100	100
1	97	97	97
2	94	94	91
3	92	92	74
4	89	89	50
5	87	85	30
10	77	50	1
20	65	6	0
30	58	1	0
40	55	0	0
50	53	0	0
100	50	0	0

The effective TPM sample is composed of particles that are collected with a TPM size-selective sampler preceded by an IPM size-selective sampler yielding the following overall fractional collection efficiency:

$$ST\ (d_{ae}) = \text{IPM}\ (d_{ae})\ [1 - F(x)] \qquad \textbf{4.4}$$

where $F(x)$ is the cumulative probability function of the standardized normal variable x, given by:

$$x = \ln\ [d_{ae}/\Gamma_t]/\ln\Sigma$$

with Γ_t = 11.64 μm and Σ =1.5 where ln is the natural logarithm. The resulting effective 50% cut size for ST(d_{ae}) is 10 μm. Selected values of the effective TPM fraction are given as a percentage in Table 4.1.

Similarly. the effective RPM sample is composed of particles that are collected with an RPM size-selective sampler preceded by an IPM size-selective sampler yielding the following overall fractional collection efficiency:

$$SR\ (d_{ae}) = \text{IPM}\ (d_{ae})\ [1 - F(x)] \qquad \textbf{4.5}$$

where $F(x)$ is the cumulative probability function of the standardized normal variable x, given by:

$$x = \ln\ [d_{ae}/\Gamma_r]/\ln\Sigma$$

with Γ_r = 4.25 μm and Σ =1.5. The resulting effective 50% cut size for $SR(d_{ae})$ is 4 μm. Selected values of the effective RPM fraction are given as a percentage in Table 4.1.

It is important to note that the collection of deliquescent or hygroscopic aerosols using either of the TPM or RPM criteria will not properly represent the actual airway penetration for these aerosols, since they will tend to grow rapidly in size in the humid environment of the respiratory tract. However, deliquescent and hygroscopic aerosols tend to be relatively soluble in body fluids so that their inhalable fraction may be of greater importance than either respirable or thoracic fractions. If respirable and/or thoracic samples are needed of hygroscopic or deliquescent aerosols, special sampling or analysis techniques may be required.

4.6 SUMMARY

The TPM fraction consists of those particles that penetrate a separator whose size collection efficiency is described by a cumulative lognormal function with median 'cut size' of d_{ae} = 11.64 μm and with geometric standard deviation 1.5, following preselection with an IPM sampler. This means that the effective overall 'cut size' is at d_{ae} = 10 μm. This criterion describes a curve for a TPM size-selective sampler which is designed for the collection and assay of those particles that penetrate the sampler.

The RPM fraction consists of those airborne particles that penetrate a separator whose size collection efficiency is described by a cumulative lognormal function with median 'cut size' of d_{ae} = 4.25 μm and with geometric standard deviation 1.5, again following preselection with an IPM sampler. This means that the effective overall 'cut size' is at d_{ae} = 4.0 μm. This criterion describes a curve for a RPM size-selective sampler which is designed for the collection and assay of those particles that penetrate the sampler.

The operation of an inhalability sampler pre-classifier adhering to the criterion presented in Chapter 3 upstream of either the respirable or thoracic selector is assumed. If this is not used, the samples will tend to somewhat overestimate the amount sampled with the respective RPM and TPM fraction. The direction of this bias, however, is such that it will provide a greater degree of protection when assessing the inhalation hazard in question.

ACKNOWLEDGMENT

The authors are grateful for the insightful contributions of Dr. Sidney C. Soderholm.

REFERENCES

Abramovitz, M. and Stegun, I.A., Eds.(1972), *Handbook of Mathematical Functions With Formulas, Graphs, and Mathematical Tables.* National Bureau of Standards Applied Mathematics Series 55, U.S. Government Printing Office, Washington, D.C.

Altshuler, B., Palmes, E.D. and Nelson, N. (1967), Regional aerosol deposition in the human respiratory tract, In: *Inhaled Particles and Vapours II* (C.N. Davies, Ed.), Pergamon Press, Oxford, pp. 323-335.

Chan, T.L. and Lippmann, M. (1980), Experimental measurements and empirical modelling of the regional deposition of inhaled particles in humans, *Am. Ind. Hyg. Assoc. J.*, 41, pp. 399-409.

Findeisen, W. (1935), Uber des absetzen kleiner, in der luft suspendierter teichen in der menschlichen lunger bei der atmung, *Pflugers Arch. Ges. Physiol.*, 236, pp. 367-379.

Foord, N., Black, A. and Walsh, M. (1976), Regional deposition of 2.5-7.5 μm diameter inhaled particles in healthy male nonsmokers, ML. 76/2892. Atomic Energy Research Establishment, Harwell, England, U.K.

George, A.C. and Breslin, A.J. (1967), Deposition of natural radon daughters in human subjects, *Health Phys.*, 13, pp. 375-378.

Giacomelli-Maltoni, G., Melandri, C., Prodi, V. *et al.* (1972), Deposition efficiency of monodisperse particles in the human respiratory tract, *Am. Ind. Hyg. Assoc. J.*, 33, pp. 603-610.

Harley, N.H., Cross, F.T. and Stuart, B.O. (1984), *Evaluation of occupational and environmental exposures to radon and radon daughters in the United States,* National Council on Radiation Protection and Measurements, NCRP Report No. 78, Bethesda, MD (1984).

Hatch, T.E. and Gross, P. (1964), *Pulmonary Deposition and Retention of Inhaled Aerosols,* Academic Press, New York.

Heyder, J. and Rudolf, G. (1977), Deposition of aerosol particles in the human nose, *Inhaled Particles IV* (W.H. Walton, Ed.), Pergamon Press, Oxford, pp.107-125.

Heyder, J., Armbruster, L., Gebhart, J. *et al.* (1975), Total deposition of aerosol particles in the human respiratory tract for nose and mouth breathing, *J. Aerosol Sci.*, 6, pp. 311-328.

Hounam, R.F. (1971), The deposition of atmospheric condensation nuclei in the nasopharyngeal region of the human respiratory tract, *Health Phys.*, 20, pp. 219-220.

Hounam, R.F., Black, A. and Walsh, M. (1971), The deposition of aerosol particles in the nasopharyngeal region of the human respiratory tract, In: *Inhaled Particles III* (W.H. Walton, Ed.), Unwin Brothers Ltd., Surrey, U.K., pp. 71-80.

International Commission on Radiological Protection (ICRP) (1994), *Human respiratory tract model for radiological protection*, ICRP Publication 66, Pergamon Press, Elmsford, NY.

International Standards Organisation (ISO) (Working Group of Committee TC 146) (1981), Recommendation on size definitions for particle sampling, *Am. Ind. Hyg. Assoc. J.*, 42, A64-A68.

Landahl, R.D. (1950), On the removal of airborne droplets by the human respiratory tract, I. The Lung, *Bull. Math. Biophys.*, 112, pp. 43-56.

Lippmann, M. (1970a), Respirable dust sampling, *Am. Ind. Hyg. Assoc. J.*, 31, pp. 138-159.

Lippmann, M.(1970b), Deposition and clearance of inhaled particles in the human nose, *Ann. Otol.*, 79, pp. 519-528.

Lippmann, M. (1977), Regional deposition of particles in the human respiratory tract, In: *Handbook of Physiology, Section 9, Reactions to Environmental Agents* (D.H.K. Lee, H.L. Falk and S.D. Murphy, Eds.), American Physiological Society, Bethesda, MD, pp. 213-232.

Lippmann, M. (1990), Effects of fiber characteristics on lung deposition, retention, and disease, *Environ. Health Perspect.*, 88, pp. 311-317

Lippmann, M. and Albert, R. (1969), The effect of particle size on the regional deposition of inhaled aerosols in the human respiratory tract, *Am. Ind. Hyg. Assoc. J.*, 30, pp. 257-275.

McClellan, R.O., Miller, F.J., Hesterberg, T.W. *et al.* (1992), Approaches to evaluating the toxicity and carcinogenicity of man-made fibers, *Reg. Tox. Pharmacol.*, 16, pp. 321-364.

Morrow, P.E., Bates, D.V., Fish, B.R. *et al.* (1966), Deposition and retention models for internal dosimetry of the human respiratory tract, Report of the International Commission on Radiological Protection, ICRP, Task Group on Lung Dynamics, *Health Phys.*, 12, pp. 173-207.

National Council on Radiation Protection and Measurements (NCRP) (1997), *Deposition, retention and dosimetry of inhaled radioactive substances*, NCRP Report No. 125, Bethesda, MD.

Pattle, R.E. (1961), The retention of gases and particles in the human nose, In: *Inhaled Particles and Vapours* (C.N. Davies, Ed.), Pergamon Press, Oxford, pp. 301-309.

Raabe, O.G. (1976), Aerosol aerodynamic size conventions for inertial sampler calibration, *J. Air Poll. Cont. Assoc.*, 26, pp. 856-860.

Raabe, O.G. (1980), Physical properties of aerosols affecting inhalation toxicology, In: *Pulmonary Toxicology of Respirable Particles* (C.L. Sanders, F.T. Cross, O.E. Dagle and J.A. Mahaffey, Eds.), CONF791002, U.S. Department of Energy, National Technical Information Service, Springfield, VA, pp. 1-28.

Raabe, O.G. (1982), Deposition and clearance of inhaled aerosols, In: *Mechanisms in Respiratory Toxicology, Vol. I* (H. Witschi and P. Nettesheim, Eds.), CRC Press, Boca Raton, FL , pp. 27-76.

Shanty, F. (1974), Deposition of ultrafine aerosols in the respiratory tract of human volunteers, Ph.D. Thesis, School of Hygiene and Public Health, The Johns Hopkins University, Baltimore, MD.

Snyder, W.S. (1975), Report of the Task Group on Reference Man, International Commission on Radiological Protection (ICRP), Pergamon Press, Oxford.

Stahlhofen, W., Gebhart, J. and Heyder, J. (1980), Experimental determination of the regional deposition of aerosol particles in the human respiratory tract, *Am. Ind. Hyg. Assoc. J.*, 41, pp. 385-398.

Stuart, B.O., Lioy, P.J. and Phalen, R.F. (1985), Use of size-selection in establishing TLVs, In: *Particle Size-Selective Sampling in the Workplace* (R.F. Phalen, Ed.), American Conference of Governmental Industrial Hygienists (ACGIH), Cincinnati, OH, pp. 65-76.

Chapter 5

SAMPLING CRITERIA FOR THE FINE FRACTIONS OF AMBIENT AIR

Morton Lippmann

Nelson Institute of Environmental Medicine, New York University School of Medicine

5.1 INTRODUCTION AND BACKGROUND

The ACGIH criteria for *inhalable*, *thoracic* and *respirable* particulate matter described in this book are all based on penetration through portals of entry: that is, a) *inhalable* for entry into the nose or mouth; b) *thoracic* for penetration through the larynx into lung airways; and c) *respirable* for penetration beyond the terminal bronchioles into the gas exchange region. This is consistent with the definitions of other standards-setting bodies, including the International Standards Organisation (ISO) and the Comité Européen de Normalisation (CEN). Another valid rationale for size-selective aerosol sampling in relation to health hazard evaluation, however, is the separation of the components of an aerosol that derive from different sources and so have different chemical properties and, possibly, different hazard potential. This approach becomes feasible when the different components have distinctly different size distributions. One notable example of the application of this use of particle size discrimination is the National Ambient Air Quality Standard for fine particulate matter promulgated by the Administrator of the U.S. Environmental Protection Agency (EPA) in July of 1997. The rationale for the selection of the 2.5-μm particle aerodynamic diameter cut point adopted in this new standard is discussed below in this chapter. The specific performance standards mandated by EPA for samplers to meet their criteria are discussed in a later chapter, along with a summary of available data on the performance of samplers used for this application of size-selective sampling.

One major responsibility of the EPA Administrator under The Clean Air Act is to periodically review the National Ambient Air Quality Standards (NAAQS) and, if necessary, to propose and promulgate new or revised standards (Lippmann, 1987). The primary NAAQS are concentration limits for specified averaging times that are intended to prevent adverse acute and chronic health effects associated with exposures to pollutants having numerous and widespread sources. Furthermore, they are intended to protect sensitive or susceptible segments of the general population with an adequate margin of safety. By contrast, ambient air pollutants from a limited number of point sources are controlled using a strategy based on source controls, using National Emission Standards for Hazardous Air Pollutants (NESHAPs).

There are currently six ambient air pollutants having specified NAAQS. These are commonly known as the criteria pollutants. They include four gaseous species: ozone (O_3); sulfur dioxide (SO_2); nitrogen dioxide (NO_2); and carbon monoxide (CO). Lead (Pb) in its various chemical forms was designated a criteria pollutant when leaded gasoline was in common use. If logic can prevail, it will someday be redesignated as a hazardous air pollutant. The sixth NAAQS is for ambient particulate matter (PM), without reference to chemical composition. In this sense, it is analogous to historic occupational exposure limits (OELs) for nuisance dusts, such as the ACGIH threshold limit values (TLVs) for particulates not otherwise classified (PNOCs). However, the toxicity of mixtures of chemicals in ambient air PM is generally greater than that of nuisance dust in the occupational setting.

Ambient air PM has long been associated with excess mortality and morbidity, as well as the nuisances of cinders in the eye and soiling of materials and surfaces. Such associations have generally been stronger than those for the ubiquitous pollutant gases that are also associated with combustion sources. This strength of association is remarkable, considering the historical reliance, by the epidemiological studies, on crude indices of ambient PM, especially black smoke (BS) and total suspended particulate matter (TSP). More recent epidemiological studies, many of them using monitoring data for thoracic particulate matter (measured as particles passing a pre-collector with a 50% 'cut' at d_{ae} = 10 µm, i.e., PM_{10}) have generally produced even stronger associations between ambient PM and mortality and morbidity. For those relatively few studies for which fine particulate indices of exposure (particles passing a pre-collector with 'cut' at d_{ae} = 2.5 µm, i.e., $PM_{2.5}$, and/or sulfate ion) have been available, the associations are generally the strongest of all (Lippmann and Thurston, 1996).

Sulfate ion ($SO_4^{=}$), most of which is formed in the atmosphere from the oxidation of the SO_2 emitted from fossil fuel combustion sources, is a major mass constituent of $PM_{2.5}$ in central and eastern North America. Some of it is associated with hydrogen ion (H^+), while most of it is associated with ammonium ion (NH_4^+) as a result of neutralization by ambient ammonia gas (NH_3) arising from anaerobic decay processes at ground level. Other major constituents of $PM_{2.5}$ include ammonium nitrate (especially in western North America) and fixed and organic carbon. Much of the fixed carbon is attributable to diesel engine exhaust, and most of the organic carbon is formed in the atmosphere in the photochemical reaction sequence that also leads to ambient air O_3.

The sulfate, nitrate, and fixed carbon components of $PM_{2.5}$ have been considered as causal factors for excess mortality and morbidity, but causal constituents have not yet been identified. Some of the suspect agents are listed in Table 5.1. Sulfate ion, which is by itself a very unlikely causal agent, may be a good surrogate index because of association in the ambient air with H^+ and peroxides. In any case, EPA staff selected $PM_{2.5}$ mass concentration as the exposure index for a new fine particle standard for ambient air (EPA, 1996a). The balance of this chapter presents some of the background and rationale for this selection. Some of it is drawn from the most recent Criteria Document and Staff Paper prepared by EPA (EPA, 1996a and b). Additional material is from a recent Workshop report by Lippmann *et al.* (1998).

If, in fact, one of the suspect agents listed in Table 5.1 (i.e., so-called "*ultrafine*" particles, defined by many as particles less than 0.1 µm in diameter) turns out to be a causal factor for mortality and morbidity, then future PM NAAQS may need to be based on particle number concentrations rather than mass concentrations.

5.2 AIR QUALITY AND EXPOSURE ASPECTS

The discussion of air quality and exposure aspects in the EPA criteria document considered: a) the chemistry and physics of atmospheric PM; b) analytical techniques for measuring PM mass, particle size and chemical composition; c) sources of ambient PM in the United States; d) temporal/spatial variability and trends in ambient

TABLE 5.1. Components of ambient air particulate matter (PM) that may account for some or all of the effects associated with exposures

Component	Evidence for Role in Effects	Doubts
Strong Acid (H^+)	Statistical associations with health effects in most recent studies for which ambient H^+ concentrations were measured.	Similar PM-associated effects observed in locations with low ambient H^+ levels.
	Coherent responses for *some* health endpoints in human and animal inhalation and *in vitro* studies at environmentally relevant doses.	Very limited data base on ambient concentrations.
Ultrafine Particles ($d \leq 0.1$ μm)	Much greater potency per unit mass in animal inhalation studies (H^+ and TiO_2 aerosols) than for same materials in larger diameter fine particle aerosols.	Absence of relevant data on responses in humans.
	Concept of "irritation signalling" in terms of number of particles per unit airway surface.	Absence of relevant data base on ambient concentrations.
Peroxides	Close association in ambient air with $SO_4^=$.	Absence of relevant data on responses in humans or animals.
	Strong oxidizing properties.	Very limited data base on ambient concentrations.

U.S. PM levels; and e) human exposure relationships (EPA, 1996b). The following summarizes the conclusions that were drawn.

CHEMISTRY AND PHYSICS OF ATMOSPHERIC PARTICLES

(i) Airborne PM is not a single pollutant, but rather is a mixture of many subclasses of pollutants with each subclass containing many different chemical species. Atmospheric PM occurs naturally as fine-mode and coarse-mode particles that, in addition to falling into different size ranges, differ in formation mechanisms, chemical composition, sources, and exposure relationships.

(ii) Fine-mode PM is derived from combustion material that has volatilized and then condensed to form primary particles or from precursor gases reacting in the atmosphere to form secondary particles. New fine-mode particles are formed by the nucleation of gas phase species and grow by coagulation (existing particles combining) or condensation (gases condensing on existing particles). Fine particles are present in two separate modes: (a) freshly generated particles, in a transient ultrafine or *nuclei* mode whose particles coagulate and move into; and (b) an *accumulation mode*, so called because its particles remain in that mode until removed by precipitation.

(iii) Coarse-mode PM, usually present by contrast in a single mode, is formed by natural processes (e.g., volcanic eruptions, fires, earthquakes, etc.) as well as by anthropogenic crushing, grinding, and abrasion of surfaces, which break large pieces of material into smaller pieces. Such particles are then suspended by the wind or by anthropogenic activity. Energy considerations limit the breakup of large particles and small particle aggregates generally to a minimum size of about 1 μm in diameter. Mining and agricultural activities are examples of anthropogenic sources of coarse-mode particles. Fungal spores, pollen, and plant and insect fragments are examples of natural bioaerosols also suspended as coarse-mode particles.

(iv) Within atmospheric particle modes, the distribution of particle number, surfaces, volume, and mass by diameter is frequently approximated by lognormal distributions (as illustrated in Figure 5.1) expressed as functions of particle aerodynamic diameter (d_{ae}). Typical values of the *mass median aerodynamic diameter* (MMAD) and geometric standard deviation (σ_g) of each size mode of an aerosol are:

Nuclei mode: *MMAD* = 0.05 to 0.07 μm; $\sigma_g = 1.8$

Accumulation mode: *MMAD* = 0.3 to 0.7 μm; $\sigma_g = 1.8$

Coarse mode: *MMAD* = 6 to 20 μm; $\sigma_g = 2.4$

At high relative humidities, or in air containing evaporating fog or cloud droplets, the accumulation mode may be split into a *droplet mode* (*MMAD* = 0.5 to 0.8 μm) and a *condensation mode* (*MMAD* = 0.2 to 0.3 μm).

(v) Research studies have used: a) impactors to determine mass as a function of size over a wide range; and b) particle counting devices to determine number as a function of particle size. Such studies have indicated an atmospheric bimodal distribution of fine and coarse particle mass with a minimum in the distribution for d_{ae} between about 1 and 3 μm. Routine monitoring studies, however, have generally been limited to measuring total suspended particles (TSP) including both fine and coarse particles of d_{ae} up to 40 or more μm; thoracic particles or PM_{10} (upper size limited by a 50% 'cut' at $d_{ae} = 10$ μm); fine particles or $PM_{2.5}$ (upper size limited by a 50% cut point at $d_{ae} = 2.5$ μm) and the coarse fraction of PM_{10} which is the difference between PM_{10} and $PM_{2.5}$ (i.e., $PM_{10-2.5}$). Cut points are not perfectly sharp for any of these PM indicators; some particles larger than the cut point are collected and some smaller-particles smaller than the cut point are not retained. These relationships are illustrated in Figure 5.1.

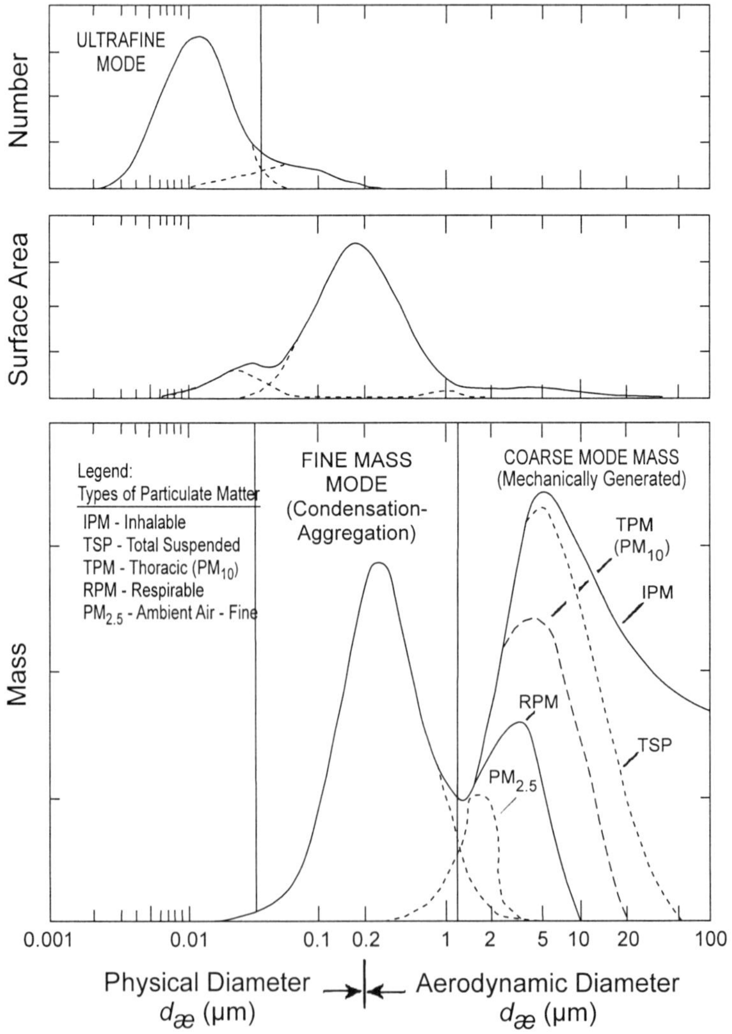

FIGURE 5.1. A hypothetical distribution of particles by number, surface area, and mass. Truncations in lower right indicate cuts imposed by aerodynamic size-selective inlets.

(vi) The terms "fine" and "coarse" were originally intended to apply to the two major atmospheric particle distributions which overlap in the size range between 1 and 3 µm. Now, "fine" has been defined by EPA as $PM_{2.5}$ and "coarse" as $PM_{10-2.5}$. However, $PM_{2.5}$ may also contain, in addition to the fine-particle mode, some of the smaller sized particles in the tail of the coarse particle mode between d_{ae} about 1 and 2.5 µm. Conversely, under high relative humidity conditions, the larger particles in the accumulation mode extend into the d_{ae} = 1 to 3 µm range. The rationale for EPA's selection of d_{ae} = 2.5 µm as the optimal cut-size for a primary $PM_{2.5}$ NAAQS will be discussed later in this chapter.

(vii) Three approaches are used to classify particles by size: a) modes, based on formation mechanisms and the modal structure observed in the atmosphere, e.g., nuclei and accumulation modes, which together comprise the fine particles; and the coarse particle mode; b) cut point, based on the 50% cut point of the specific sampling device, e.g., $PM_{2.5}$, $PM_{10-2.5}$, and PM_{10}; and c) dosimetry, based on the ability of particles to enter certain regions of the respiratory tract.

SOURCES OF AIRBORNE PARTICLES IN THE UNITED STATES

(i) The chemical complexity of airborne particles requires that the composition and sources of a large number of primary and secondary components be considered. Major components of fine particles are $SO_4^{=}$, H^+, NO_3^-, NH_4^+, organic compounds, trace elements (including metals that volatize at combustion temperatures), elemental carbon and water. Major sources of these fine mode substances are fossil fuel combustion by electric utilities, industry and motor vehicles; vegetation burning; and the smelting or other processing of metals.

(ii) Sulfur dioxide (SO_2), nitrogen oxides (NO_x), and certain organic compounds are major precursors of fine secondary PM. NO reacts with ozone (O_3) to form NO_2. SO_2 and NO_2 react with hydroxyl radical (OH) during the daytime to form sulfuric and nitric acid. During the night-time, NO_2 reacts with O_3 and forms nitric acid (HNO_3) through a sequence of reactions involving the nitrate radical ($N^{\bullet}O_3$). These acids may react further with NH_3 to form ammonium sulfate and nitrate. Some types of higher molecular weight organic compounds react with OH radicals, and olefinic compounds also react with O_3 to form oxygenated organic compounds which can condense onto existing particles.

SO_2 also dissolves in cloud and fog droplets where it may react with dissolved O_3, H_2O_2, or, if catalyzed by certain metals, with O_2, yielding sulfuric acid or sulfates, that lead to $PM_{2.5}$ when the droplet evaporates.

(iii) The formation of secondary particles depends on reactions involving O, O_3, and H_2O_2, species which are normally present in the atmosphere, but which are generated in higher concentrations during the photochemical smog formation process. Since smog formation increases with sunlight and temperature, secondary PM peaks during the summer in most U.S. areas.

(iv) Background geogenic and biogenic emission sources include: a) windblown dust from erosion and reentrainment; b) long-range transport of dust from the Sahara desert; c) sea salt; d) particles formed from the oxidation of sulfur compounds emitted from oceans and wetlands; the oxidation of NO_x from natural forest fires and lightning; and e) the oxidation of hydrocarbons (such as terpenes) emitted by vegetation.

(v) Major components of coarse particles are aluminosilicates and other oxides of crustal elements (e.g., Fe, Ca, etc.) in soil dust; fugitive dust from roads, industry, agriculture, construction and demolition; fly ash from combustion of oil and coal; and additional contributions from plant and animal material.

(vi) Fugitive dust constitutes about 90% of estimated PM_{10} emissions in the United States. Emissions are sporadic and widespread. Only a small percentage of this material is emitted in the fine particle size fraction.

(vii) Uncertainties in emissions inventory estimates could range from about 10% for well defined sources (e.g., SO_2) to an order of magnitude for widespread and sporadic sources (e.g., fugitive dust).

(viii) There was no clear trend in estimated emissions of fugitive dust and emissions from natural sources between 1984 and 1993. Estimated primary PM_{10} emissions from combustion sources decreased by about 10%; estimated SO_2 emissions decreased by about 6%; and there was no significant change in estimated NO_x emissions.

(ix) Receptor modeling has proven to be a useful method for identifying contributions of different types of sources especially for the primary components of ambient PM. Apportionment of secondary PM is more difficult because it requires consideration of atmospheric reaction processes and rates. Results from western U.S. sites indicate that fugitive dust, motor vehicles, and wood smoke are the major contributors to ambient PM samples there, while results from eastern U.S. sites indicate that stationary combustion and fugitive dust are major contributors to ambient PM samples in the East. Sulfate and organic carbon are the major secondary components in the East, while nitrates and organic carbon are the major secondary components in the West.

(x) Fine and coarse particles have distinctly different sources, both natural and anthropogenic. Therefore, different control strategies are likely to be needed.

ATMOSPHERIC TRANSPORT AND FATE OF AIRBORNE PARTICLES

(i) Dry deposition of fine particles is slow. Nuclei-mode (ultrafine) particles are rapidly removed by coagulation into accumulation-mode particles. Accumulation-mode particles are removed from the atmosphere primarily by forming cloud droplets and falling out in raindrops. Coarse particles are removed mainly by gravitational settling and inertial impaction.

(ii) Primary and secondary fine particles have long lifetimes in the atmosphere (days to weeks) and travel long distances (hundreds to thousands of kilometers). They tend to be uniformly distributed over urban areas and larger regions, especially in the eastern United States. As a result, they are not easily traced back to their individual sources.

(iii) Coarse particles normally have shorter lifetimes (minutes to hours) and only travel short distances (< 10's of km). Therefore, coarse particles tend to be unevenly distributed across urban areas and tend to have more localized effects than fine particles (although dust storms occasionally cause long-range transport of the smaller coarse-mode particles.)

5.3 CHARACTERIZATION OF AMBIENT PM CONCENTRATIONS

Table 5.2 introduces some of the size-related terminology used by EPA in its criteria document (EPA, 1996b) and staff paper (EPA, 1996a). PM_x (where $x \equiv 1, 2.5, 10, 15, 10\text{–}2.5$) is used to refer to gravimetric measurements with a 50% 'cut' of x µm in particle aerodynamic diameter, while the terms "fine" or "coarse" particles will be used more generally to refer to the fine and coarse modes of the particle distribution. The distinction highlights the role of formation mechanism and chemistry in addition to size in defining fine and

TABLE 5.2. Particle size fraction terminology used in EPA staff paper (1996b)

Term	Description
Size Distribution Modes	
Fine particles	Fine mode particles, which are generally formed through chemical reaction, nucleation, condensation of gases, and coagulation of smaller particles; contains most numerous particles and represents most surface area.
Coarse Particles	Coarse mode particles, which are mostly generated from mechanical processes through crushing or grinding.
Sampling Measurements	
Total Suspended Particles	Particle measured by a high volume sampler as described in 40 CFR Part 50, Appendix B. This sampler has a cut point of aerodynamic diameters* that varies between 25 and 40 μm, depending on wind speed and direction.
PM_{10}	Particles measured by a sampler that contains a size fractionator (classifier) designed to have an effective cut point of 10 μm aerodynamic diameter. This measurement includes the fine mode and part of the general coarse mode and is an indicator for thoracic particles (i.e., particles that penetrate to the tracheobronchial and the gas-exchange regions of the lung).
$PM_{2.5}$	Particles measured by a sampler that contains a size fractionator (classifier) designed to have an effective cut point of 2.5 μm. The collected particles include most of the fine mode. Some small portion of the coarse mode may be included depending on the sharpness of the sampler efficiency curve and the size of coarse mode particles present.
Coarse fraction of PM_{10}, i.e., $PM_{(10-2.5)}$	Particles measured directly using a dichotomous sampler or by subtraction of particles measured by a $PM_{2.5}$ sampler from those measured by a PM_{10} sampler.

*When discussing samplers, cut point is a term used to describe the separation efficiency curve for particle collection devices. The cut point is typically described by the particular aerodynamic diameter at which the sampler achieves 50% collection efficiency. Aerodynamic diameter is defined as the diameter of a spherical particle with equal settling velocity but a material density of 1 g/cm^3. This normalizes particles of different shapes and densities.

coarse mode particles. Any specific measurement (e.g., $PM_{2.5}$) is only an approximation for fine particles.

In addition to gravimetric fine particle measurements, PM has been characterized in the U.S. and abroad using a variety of filter-based optical techniques including British or black smoke (BS), coefficient of haze (COH), and carbonaceous material (KM), as well as estimates derived from visibility measurements. In locations where they are calibrated to standard mass units (e.g., London), these measurements can be useful as surrogates for fine particle mass.

The distinction between any specific measurement of fine particles and fine mode (or a measurement of coarse particles and coarse mode) is important because the public health consequences differ between fine and coarse mode particles. Examples of fine particle measurements include $PM_{2.5}$, BS, COH and concentrations of specific chemical classes predominantly in the fine fraction such as $SO_4^=$ and H^+ can all be considered to be surrogates for fine mode particles. Measurements of coarse particles include $PM_{10-2.5}$, $PM_{15-2.5}$, and (TSP minus PM_{10}). As summarized in Table 5.3, fine and coarse particles can be differentiated by their sources and formation processes, chemical composition, solubility, acidity, atmospheric lifetime and behavior, and transport distances.

5.4 Rationales for particle size-selective sampling of ambient Aerosol

In 1987, EPA replaced the PM NAAQS based on total suspended particulate matter (TSP) with a criterion based on thoracic particulate matter, defined as PM_{10} . In 1997, following its most recent thorough review of the literature on the health effects of ambient PM, EPA concluded that most of the health effects attributable to PM in ambient air were more closely associated with the fine particles in the accumulation mode than with the coarse mode particles within PM_{10}. The Agency promulgated new NAAQS based on fine particles, defined as $PM_{2.5}$, to supplement the PM_{10} NAAQS that was retained (EPA, 1996b). The selection of $d_{ae} = 2.5$ µm as the criterion for defining the upper bound of fine particles in a regulatory sense was, inevitably, an arbitrary selection made from a range of possible options. It was arrived at using the following rationales:

(i) Fine particles produce adverse health effects because of their chemical composition (see Tables 5.2 and 5.3) more than their size, and need to be regulated using an index which is responsive to control measures being applied to direct and indirect sources of such particles.

(ii) Any separation by aerodynamic particle size that attempts to separate fine mode from coarse mode particles cannot include all fine mode particles and exclude all coarse mode particles because the modes overlap (see Figure 5.1).

(iii) The position of the 'saddle point' between the fine mode and coarse mode peaks varies with aerosol composition and climate. Figure 5.1, based on data from Michigan, indicates a volumetric

TABLE 5.3. Comparison of ambient fine and coarse mode particles*

	Fine Mode	Coarse Mode
Formed from:	Gases	Large solids/droplets
Formed by:	Chemical reaction; nucleation; condensation; coagulation; evaporation of fog and cloud droplets in which gases have dissolved and reacted.	Mechanical disruption (e.g., crushing, grinding, abrasion of surfaces); evaporation of sprays; suspension of dusts.
Composed of:	Sulfate, $SO_4^{=}$; nitrate, NO_3^-; ammonium, NH_4^+; hydrogen ion, H^+; elemental carbon; organic compounds (e.g., PAHs, PNAs); metals (e.g., Pb, Cd, V, Ni, Cu, Zn, Mn, Fe); particle-bound water.	Resuspended dusts (e.g., soil dust, street dust); coal and oil fly ash; metal oxides of crustal elements (Si, Al, Ti, Fe); $CaCO_3$, NaCl, sea salt, pollen, mold spores; plant/animal fragments; tire wear debris.
Solubility:	Largely soluble, hygroscopic and deliquescent.	Largely insoluble and nonhygroscopic.
Sources:	Combustion of coal, oil, gasoline, diesel, wood; atmospheric transformation products of NO_x, SO_2, and organic compounds including biogenic species (e.g., terpenes); high temperature processes, smelters, steel mills, etc.	Resuspension of industrial dust and soil tracked onto roads; suspension from disturbed soil (e.g., farming, mining, unpaved roads); biological sources; construction and demolition; coal and oil combustion; ocean spray.
Lifetimes:	Days to weeks	Minutes to hours
Travel Distance:	100s to 1000s of kilometers	< 1 to 10s of kilometers

*Source: EPA (1996b)

saddle point at $d_{ae} \sim 2$ µm. If the data were corrected for particle density, it might be somewhat higher.

(iv) Preferences expressed for the optimal cut size are regionally biased. In the humid eastern U.S., where the concentrations of hygroscopic $SO_4^{=}$ aerosols are highest, where the droplets are relatively large, and where windblown soil is a relatively small contributor to fine particle mass, the preference is for the upper end of the range (i.e., $PM_{2.5}$) to ensure that the sample includes essentially all of the $SO_4^{=}$. In the more arid west, where the fine mode contains much less hygroscopic aerosol, and where there can be considerable mass of coarse mode particles below $d_{ae} = 2.5$ µm, the preference expressed was for a 'cut' at or near $d_{ae} = 1$ µm.

(v) Evidence for a need for a fine particle NAAQS came from studies based on $PM_{2.5}$ or $PM_{2.1}$. Also, $PM_{2.5}$, if it errs, does so on the conservative side with respect to health protection. Further, it was deemed to be impractical to have different cut-sizes in different parts of the U.S.

(vi) The intrusion of coarse mode mass into $PM_{2.5}$ can be minimized by specifying a relatively sharp cut-characteristic for the $PM_{2.5}$ reference sampler (i.e., $\sigma_g = 1.5$).

5.5 Evidence for Health Effects of Fine Particles

The epidemiological evidence for health effects attributable to exposures to fine particles ($PM_{2.5}$) of outdoor origin is more limited than for thoracic particles (PM_{10}), primarily due to the paucity of $PM_{2.5}$ monitoring data. However, the evidence that is available enables a reasonable argument to be made about the role of fine particles in relation to health effects. In particular, since $PM_{2.5}$ is a major component of PM_{10}, typically 60–70% in the eastern U.S. and 35–40% in the western U.S., it may also account for much of the frequently reported association between health effects and PM_{10}.

One major study for which fine particles ($PM_{2.1}$) were measured, as well as thoracic particles and TSP, on a daily basis was the Harvard Six-Cities Study. Results of daily time-series mortality analyses from this study (Schwartz *et al.*, 1996), as replotted in the EPA criteria document (EPA, 1996b), are illustrated in Figure 5.2. It can be seen here that both fine particle mass and its sulfate component are highly correlated with daily mortality, but that the PM indices representing or including coarser fractions of the ambient PM are much less closely associated.

Annual average fine particle concentrations in ambient air have been significantly related to variations in annual mortality rates in cohort studies that accounted for known influences of personal risk factors such as diet, lifestyle and cigarette smoking (Dockery *et al.*, 1993; Pope *et al.*, 1995). Some of the results from these studies are presented in Table 5.4. It can be seen that the excesses in all cause deaths are accounted for by the larger excesses in cardiopulmonary and lung cancer deaths and that deaths from all other causes are not related to city-to-city variations in PM pollution.

Small, but statistically significant variations in lung capacity and bronchitis in children have also been associated with annual average fine particle concentrations in 22 U.S. and Canadian cities (Raizenne *et al.*, 1996; Dockery *et al.*, 1996). As illustrated in Figure 5.3, there is no comparable association when the coarse particle components of thoracic particles are regressed against reduced lung capacity.

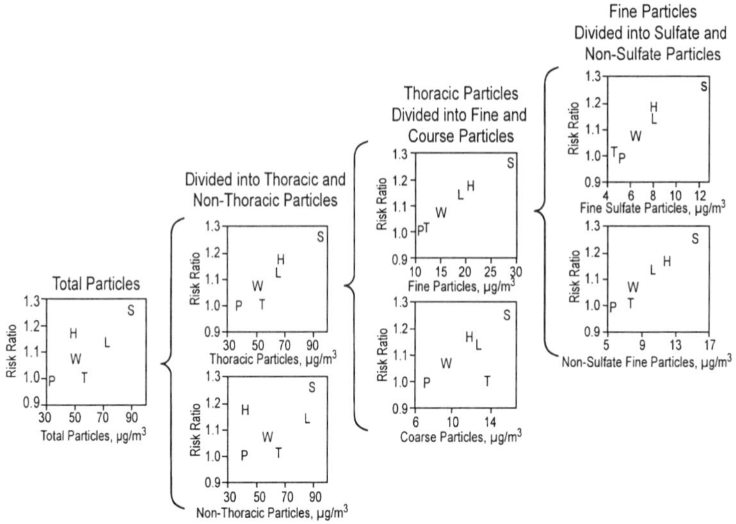

FIGURE 5.2. Adjusted relative risks for annual mortality are plotted against each of seven long-term average particle indices in the Six City Study, from largest size range, total suspended particulate matter (lower left), through sulfate and non-sulfate fine particle concentrations (upper right). Note that a relatively strong linear relationship is seen for fine particles and for its sulfate and non-sulfate components. Topeka , KS (T), which has a substantial coarse particle component of thoracic particle mass, stands apart from the linear relationship between relative risk and thoracic particle concentration. The other cities are Steubenville, OH (S), St. Louis, MO (L), Kingston and Harriman, TN (H), Watertown, MA (W), and Portage, WI (P).

TABLE 5.4. Adjusted mortality risk ratios for smoking and for particulate matter exposure by causes of death in two recent cohort studies

	Current Smokers Versus Non-Smokers		Most Versus Least Polluted City*		
			ACS		6-City
Cause of Death	ACS	6-City	Sulfate	$PM_{2.5}$	$PM_{2.5}$
All causes	2.07 (1.75, 2.43)	2.00 (1.51, 2.65)	1.15 (1.09, 1.22)	1.17 (1.09, 1.26)	1.26 (1.08, 1.47)
Lung Cancer	9.73 (5.96, 15.9)	8.00 (2.97, 21.6)	1.36 (1.11, 1.66)	1.03 (0.50, 1.33)	1.37 (0.81, 2.31)
Cardiopulmonary	2.28 (1.79, 2.91)	2.30 (1.56, 3.41)	1.26 (1.16, 1.37)	1.31 (1.17, 1.46)	1.37 (1.11, 1.68)
All Other	1.54 (1.19, 1.99)	1.46 (0.89, 2.39)	1.01 (0.92, 1.11)	1.07 (0.92, 1.24)	1.01 (0.79, 1.30)

*ACS sulfates, 151 cities (Great Falls, MT versus Steubenville, OH); ACS fine particles, 50 cities, (Albuquerque, NM versus Huntington, WV); Six City study (Portage, WI versus Steubenville, OH).

Results of the above cited studies, and from studies of rates of bronchitis in relation to various measures of ambient PM concentrations, are shown in Table 5.5. This table also demonstrates that these effects have been seen in studies where none of the PM NAAQS (for TSP or PM_{10}) have been exceeded.

Summaries from the recent EPA staff paper (EPA, 1996a) of the minimum levels of $PM_{2.5}$ that have been indicated as levels associated with health effects by the EPA are shown in Tables 5.6 and 5.7 for 24-hour and annual average concentrations, respectively.

5.6 CONCERN ABOUT HEALTH EFFECTS OF COARSE MODE PM_{10}

While mortality, morbidity and lung function decrements have been more closely associated with fine particle mass concentrations than with coarse particle mass, it is not clear that fine particle NAAQS alone would provide adequate public health protection. The coarse mode particles within the thoracic fraction (as defined by PM_{10}) deposit preferentially in the lung conductive airways where they may contribute to asthma exacerbations and to the development and/or progression of chronic bronchitis and bronchial cancer. Further-

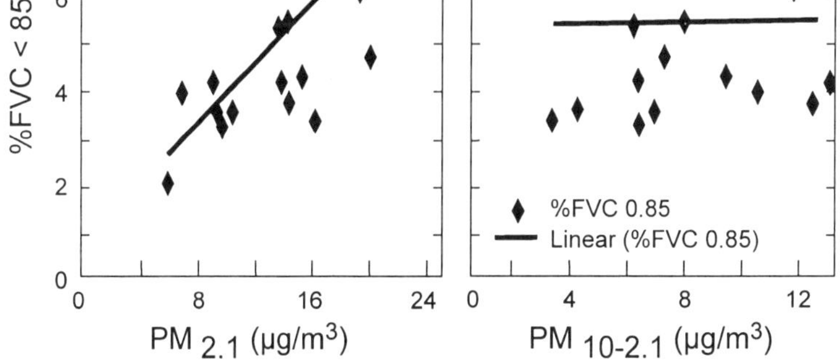

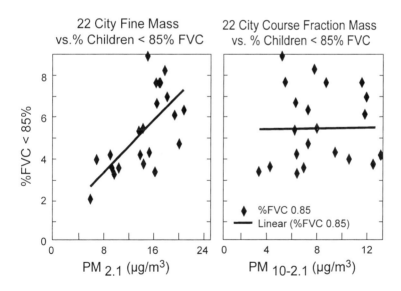

FIGURE 5.3. Percentage of children with < 85% normal FVC vs. annual fine and coarse fraction mass in 22 City study. (EPA graphical depiction of results from Raizenne *et al.*, 1996; Spengler *et al.*, 1996). The relationship between fine mass and lung function decrement is significant. No clear relation is shown for coarse fraction particles, which are generally at low concentration in these cities.

TABLE 5.5. Effect estimates per increments* in annual mean levels of fine particle indicators from U.S. and Canadian studies

Type of Health Effect & Location	Indicator	Change in Health Indicator per Increment in PM*	Range of City PM Levels Means ($\mu g/m^3$)
Increased Total Annual Mortality in Adults		Relative Risk (95% CI)	
Six City (Dockery *et al.* 1993)	$PM_{15/10}$	1.42 (1.16–2.01)	18–47
	$PM_{2.5}$	1.31 (1.11–1.68)	11–30
	$SO_4^=$	1.46 (1.16–2.16)	5–13
ACS Study (Pope *et al.* 1995)	$PM_{2.5}$	1.17 (1.09–1.26)	9–34
(151 U.S. SMSA)	$SO_4^=$	1.10 (1.06–1.16)	4–24
Increased Bronchitis in Children		Odds Ratio (95% CI)	
Six City (Dockery *et al.* 1989)	$PM_{15/10}$	3.26 (1.13, 10.28)	20–59
Six City (Ware *et al.* 1986)	TSP	2.80 (1.17, 7.03)	39–114
24 City (Dockery *et al.* 1996)	H^+	2.65 (1.22, 5.74)	6.2–41.0
24 City (Dockery *et al.* 1996)	$SO_4^=$	3.02 (1.28, 7.03)	18.1–67.3
24 City (Dockery *et al.* 1996)	$PM_{2.1}$	1.97 (0.85, 4.51)	9.1–17.3
24 City (Dockery *et al.* 1996)	PM_{10}	3.29 (0.81, 13.62)	22.0–28.6
Southern California (Abbey *et al.* 1995)	$SO_4^=$	1.39 (0.99, 1.92)	—
Decreased Lung Function in Children			
Six City (Dockery *et al.* 1989)	$PM_{15/10}$	NS⁺	20–59
Six City (Ware *et al.* 1986)	TSP	NS⁺	39–114
24 City (Raizenne *et al.* 1996)‡	H^+ (52 nmols/m^3)	−3.45% (−4.87, −2.01) FVC	—
24 City (Raizenne *et al.* 1996)‡	$PM_{2.1}$ (15 $\mu g/m^3$)	−3.21% (−4.98, −1.41) FVC	—
24 City (Raizenne *et al.* 1996)‡	$SO_4^=$ (7 $\mu g/m^3$)	−3.06% (−4.50, −1.60) FVC	—
24 City (Raizenne *et al.* 1996)‡	PM_{10} (17 $\mu g/m^3$)	−2.42% (−4.30, −0.51) FVC	—

* Annual-average PM increments assume: 100 $\mu g/m^3$ TSP; 50 $\mu g/m^3$ PM_{10} and PM_{15}; 25 $\mu g/m^3$ $PM_{2.5}$; 100 nmole/m^3 H^+; and 15 $\mu g/m^3$ $SO_4^=$ (except where noted otherwise).

⁺NS = No signficant changes.

‡ Pollutant data same as for Dockery *et al.* (1996).

TABLE 5.6. Estimated levels of minimum clear increased risk in terms of measured or estimated $PM_{2.5}$ (24-hour average)

Study/PM Indicator	Minimum Clear Increased Risk Level for PM (μg/m³ of measured or estimated $PM_{2.5}$)
MORTALITY – 6 Cities – $PM_{2.5}$	
Schwartz *et al.* (1996)	< 30
MORTALITY – St. Louis - $PM_{2.5}$/PM_{10}	
Dockery *et al.* (1992)	21[1]
Samet *et al.* (1995)	28[1]
MORTALITY – Utah Valley – PM_{10}	
Pope *et al.* (1992)	27[2]
Samet *et al.* (1995)	29[2]
MORTALITY – Birmingham – PM_{10}	
Schwartz (1993a)	34[3]
Samet *et al.* (1995)	30[3]
MORTALITY – Philadelphia – TSP	
Samet *et al.* (1995)	34[4]
HOSPITAL ADMISSIONS – Birmingham – PM_{10}	
Schwartz (1994a)	11–26[3]
HOSPITAL ADMISSIONS – Detroit – PM_{10}	
Schwartz (1994b)	33[5]
HOSPITAL ADMISSIONS (Cardiac) – Detroit – PM_{10}	
Schwartz and Morris (1995)	23[5]
HOSPITAL ADMISSIONS – Ontario – Sulfate	
Burnett *et al.* (1994 and 1995)	13–18[6]
RESPIRATORY SYMPTOMS – 6 Cities – $PM_{2.5}$	
Schwartz *et al.* (1994)	≤ 18[7]
RESPIRATORY SYMPTOMS – Utah Valley – PM_{10}	
Pope and Dockery (1992)	32[2]

[1]Concentration to $PM_{2.5}$ from PM_{10} quartile done by using a site-specific $PM_{2.5}$/PM_{10} ratio for the period of study (0.64).

[2]Conversion to PM_{10} used nationwide $PM_{2.5}$/PM_{10} ratio for all seasons (0.58) from SAI (1996), because urbanized Utah Valley judged not well represented by other Southwest sites.

[3]Minimum clear risk indicates lowest concentration on nonparametric smoothed curve where a clear and consistent increase in risk is evident. Conversion to $PM_{2.5}$ done using the $PM_{2.5}$/PM_{10} ratio for Southeast region from SAI (1996) (0.57).

[4]Conversion to $PM_{2.5}$ done by applying median $PM_{2.5}$/TSP ratio available for 1982 in Philadelphia (0.34), using data from the inhalable particle network.

[5]Conversion to $PM_{2.5}$ done using $PM_{2.5}$/PM_{10} ratio (0.63) for central U.S. from SAI (1996).

[6]Conversion to $PM_{2.5}$ done using site-specific regression equations to convert from sulfate to PM_{10} for the three major cities in the study and converting the PM_{10} values to $PM_{2.5}$ using a nationwide $PM_{2.5}$/PM_{10} regression equation for Canada (Brook *et al.*, 1997). The results agree closely with those obtained using a sulfate/$PM_{2.5}$ ratio for Toronto from Thurston *et al.* (1994).

[7]Conversion to $PM_{2.5}$ from PM_{10} done by using site-specific ratio for the period of study (0.6).

TABLE 5.7. Estimated levels of minimum clear increased risk in terms of measured or estimated $PM_{2.5}$ (annual average)

Study/PM Indicator	Minimum Clear Increased Risk Level for PM ($\mu g/m^3$ of measured or estimated $PM_{2.5}$)
MORTALITY – 6 Cities – $PM_{2.5}$ Dockery *et al.* (1993)	15 – <30
MORTALITY – ACS 50 Cities – $PM_{2.5}$ Pope *et al.* (1995)	> 15
BRONCHITIS in CHILDREN – 6 Cities – $PM_{2.5}/PM_{10}$ Dockery *et al.* (1989)	22
CHRONIC BRONCHITIS – 53 Cities – TSP Schwartz (1993b)	23*

*Derived from applying conversion factors for PM_{10}/TSP (0.5) and $PM_{2.5}/PM_{10}$ (0.6) for late 1970s – early 1980s data and previous. Ratios may have differed somewhat over the study period (1970–1974).

more, even if the effects are due to the particles that deposit in the gas exchange region of the lungs, there may be high concentrations of respirable coarse mode particles; that is, particles with aerodynamic diameters below the RPM 50% 'cut' of d_{ae} = 4 µm and the fine mode 'cut' of d_{ae} = 2.5 µm, in the arid portions of the western U.S. on windy days.

5.7 Establishment of PM NAAQS for Multiple Particle Size Fractions

In consideration of current knowledge of the health effects of ambient air PM, as summarized above, the EPA Administrator, in July of 1997, established new daily and annual fine particle ($PM_{2.5}$) NAAQS and retained the daily and annual thoracic (PM_{10}) NAAQS (EPA, 1996a). The alternate option of specific thoracic coarse mode NAAQS based on the mass concentration of PM_{10} in particles larger than d_{ae} = 2.5 µm was rejected as being too complex to be implemented. The rationale was that, with the existence of a $PM_{2.5}$ NAAQS, the PM_{10} NAAQS would serve adequately to limit excessive exposures to coarse mode thoracic particles. With this approach, regulated communities can continue to use their existing PM_{10} monitors to supplement their newly-installed $PM_{2.5}$ monitors. They can also monitor $PM_{2.5}$ and PM_{10} simultaneously using one instrument, such as a virtual dichotomous sampler with a 10-µm inlet and a 2.5-µm separator, provided that the instrument selected can be demonstrated to provide equivalent performance to the established Federal Reference Methods for both PM_{10} and $PM_{2.5}$.

5.8 Discussion

In considering whether the epidemiological evidence cited above was sufficient for the establishment of more restrictive PM NAAQS, the EPA Administrator was handicapped by the absence of clear evidence from controlled exposure studies in humans or laboratory animals using PM components and mixtures that similar or related effects occur, as well as from a lack of established biological mechanisms that could account for adverse effects such as excess mortality and hospital admissions for cardiopulmonary diseases. In a sense, the absence of such effects in controlled human studies is consistent with the epidemiological evidence. The only documented effects in healthy humans from peak daily exposures to ambient air pollutants are relatively small changes in respiratory function and symptoms, and these are more closely associated with O_3 than with PM.

The primary PM standards are designed to protect sensitive subgroups of the overall population from adverse health effects. The relevant subgroups include infants and the elderly, especially those with pre-existing respiratory diseases such as asthma, pneumonia, chronic obstructive pulmonary disease (COPD), and cardiovascular disease. Some of these people may be older workers or retired workers who have had long-term exposures to dusts and irritant vapors. Such workers are likely to constitute a susceptible sub-population. It is well established that long-term exposure to mineral dusts can lead to chronic bronchitis and accelerated loss of lung function and that these conditions can progress even after the occupational exposures end. Miners and other workers in dusty trades may retire with lung function in the normal or near normal range and then become progressively disabled. This is because of dust that accumulates around small airways, and the loss of lung recoil, which is a normal part of lung aging, results in airflow obstruction. They would be a sensitive population group. Registries of such workers, if enrolled in prospective cohort studies, may be ideal populations for further studies of both acute and chronic air pollution mortality and morbidity.

In designing future studies of the health effects of occupational exposures to PM or to mixtures of PM and irritant vapors, consideration should be given to the hypotheses generated by the recent epidemiological research on ambient air PM. Attention should be paid to cardiovascular and respiratory symptoms and functions in occupationally exposed workers, as well as to daily variations in their responses that may be related to variations in their exposures to environmental pollutants.

In summary, it is clear that the precautionary nature of the Clean Air Act has led the EPA Administrator to promulgate new $PM_{2.5}$ NAAQS on the basis of a less comprehensive data base than is desirable. Future health effects research will be needed to establish whether such standards lead to the predicted health benefits and whether further refinements and/or alternate PM standards are needed. In particular, research is needed to more firmly establish the roles played by exposure to thoracic coarse-mode particles, accumulation mode-fine particles, and ultrafine particles in the mortality and morbidity shown to be associated in past studies with PM_{10} and $PM_{2.5}$, and the optimal PM NAAQS for protecting the public health against such exposures.

REFERENCES

Abbey, D.E., Ostro, B.E., Peterson, F., and Burchette, R.J. (1995), Chronic respiratory symptoms associated with estimated long-term ambient concentrations of fine particulates less than 2.5 μm in aerodynamic diameter and other air pollutants, *J. Exp. Environ. Epidemiol.*, 5, pp. 137-159.

Brook, J.R., Dann, T.F., and Burnett, R.T. (1997), The relationship among TSP, PM_{10}, $PM_{2.5}$ and inorganic constituents of atmospheric particulate matter at multiple Canadian locations, *J. Air Waste Manage. Assoc.*, 47, pp. 2-19.

Burnett, R.T., Dales, R.E., Raizenne, M.E. *et al.* (1994), Effects of low ambient levels of ozone and sulfates on the frequency of respiratory admissions to Ontario hospitals, *Environ. Res.*, 65, pp. 172-194.

Burnett, R.T., Dales, R.E., Raizenne, M.E. *et al.* (1995), Associations between ambient particulate sulfate and admissions to Ontario hospitals for cardiac and respiratory diseases, *Am. J. Epidemiol.*, 142, pp. 15-22.

Dockery, D.W., Speizer, F.E., Stram, D.O. *et al.* (1989), Effects of inhalable particles on respiratory health of children, *Am. Rev. Respir. Dis.*, 139, pp. 587-594.

Dockery, D.W., Schwartz, J., and Spengler, J.D. (1992), Air pollution and daily mortality: Associations with particulates and acid aerosols, *Environ. Res.*, 59, pp. 362-373.

Dockery, D.W., Pope, C.A. III, Xu, X. *et al.* (1993), An association between air pollution and mortality in six U.S. cities, *N. Engl. J. Med.*, 329, pp. 1753-1759.

Dockery, D.W., Cunningham, J., Damokosh, A.I. *et al.* (1996), Health effects of acid aerosols on North American children: respiratory symptoms, *Environ. Health Perspect.*, 104, pp. 500-505.

Lippmann, M. (1987), Role of science advisory groups in establishing standards for ambient air pollutants, *Aerosol Sci. Tech.*, 6, pp. 93-114.

Lippmann, M. and Thurston, G.D. (1996), Sulfate concentrations as an indicator of ambient particulate matter air pollution for health risk evaluations, *J. Exposure Anal. Environ. Epidemiol.*, 6, pp. 123-146.

Lippmann, M., Bachmann, J.D., Bates, D.V. *et al.* (1998), Report of the Particulate Matter (PM) Research Strategies Workshop, Park City, UT, *Appl. Occup. Environ. Hyg.*, 13(6):485-493.

Pope, C.A. III and Dockery, D.W. (1992), Acute health effects of PM_{10} pollution in Utah Valley, *Arch. Environ. Health*, 47, pp. 211-217.

Pope, C.A. III, Schwartz, J. and Ransom, M.R. (1992), Daily mortality and PM_{10} pollution in Utah Valley, *Arch. Environ. Health*, 47, pp. 211-217.

Pope, C.A. III, Thun, M.J., Namboodiri, M. *et al.* (1995), Particulate air pollution is a predictor of mortality in a prospective study of U.S. adults, *Am. J. Respir. Crit. Care Med.*, 151, pp. 669-674.

Raizenne, M., Neas, L.M., Damokosh, A.I. *et al.* (1996), Health effects of acid aerosols on North American chidren: pulmonary function, *Environ. Health Perspect.*, 104, pp. 506-514.

SAI (1996), Statistical Support for Particulate Matter NAAQS, Systems International, San Rafael, CA 94903.

Samet, J.M., Zeger, S.L., and Berhane, K. (1995), The association of mortality and particulate air pollution, in: Particulate Air Pollution and Daily Mortality: Replication and Validation of Selected Studies, Health Effects Institute, Cambridge, MA (August 1995).

Schwartz, J. (1993a), Air pollution and daily mortality in Birmingham, AL, *Am. J. Epidemiol.*, 137, pp. 1136-1147.

Schwartz, J. (1993b), Particulate air pollution and chronic repiratory disease, *Environ. Res.*, 62, pp. 7-13.

Schwartz, J. (1994a), Air pollution and hospital admissions for elderly in Birmingham, AL, *Am. J. Epidemiol.*, 139, pp. 589-598.

Schwartz, J. (1994b), Air pollution and hospital admissions for the elderly in Detroit, MI, *Am. J. Respir. Crit. Care Med.*, 150, pp. 648-655.

Schwartz, J. and Morris, R. (1995), Air pollution and hospital admissions for cardiovasular disease in Detroit, MI, *Am. J. Epidemiol.*, 142, pp. 927-939.

Schwartz, J., Dockery, D.W., Neas, L.M. *et al.* (1994), Acute effects of summer air pollution on respiratory symptom reporting in children, *Am. J. Respir. Crit. Care Med.*, 150, pp. 1234-1242.

Schwartz, J., Dockery, D.W. and Neas, L.M. (1996), Is daily mortality associated significantly with fine particles? *J. Air Waste Manage. Assoc.*, 46:927-939.

Thurston, G.D., Ito, K., Hayes, C.G. *et al.* (1994), Respiratory hospital admissions and summertime haze air pollution in Toronto, Ontario: Consideration of the role of acid aerosols, *Environ. Res.*, 65, pp. 271-290.

United States Environmental Protection Agency (EPA) (1996a), *Air Quality Criteria for Particulate Matter*, EPA/600/P-95/001, Washington, DC.

United States Environmental Protection Agency (EPA) (1996b), *Review of the National Ambient Air Quality Standards for Particulate Matter*, Office of Air Quality Planning and Standards (OAQPS) Staff Paper, EPA-452/R-96-013, Environmental Protection Agency, Research Triangle Park, NC (July 1996).

Ware, J.H., Ferris, B.G. Jr., Dockery, D.W. *et al.* (1986), Effects of ambient sulfur oxides and suspended particles on respiratory health of preadolescent children, *Am. Rev. Respir. Dis.*, 133, pp. 834-842.

◆ ◆ ◆ ◆ ◆

Chapter 6

SAMPLING FOR INHALABLE AEROSOL

William C. Hinds

Department of Environmental Health Sciences, Center for Occupational and Environmental Health, School of Public Health, University of California Los Angeles.

6.1 INTRODUCTION

Proper estimation of inhalation exposure to airborne particles requires an understanding of: a) particle inhalation efficiency versus particle size for humans at work (inhalability); and b) sampling procedures that will either mimic particle inhalation efficiency for humans or permit estimation of the mass concentration of inhalable particles in the work environment. Our understanding of both of these aspects is limited. The first aspect is discussed in Chapter 3. This section deals only with the second aspect, inhalable aerosol sampling. It includes a review of previous work on inhalable sampling and a discussion of feasible approaches to inhalable sampling.

It is intended that *inhalable particulate matter* (IPM) sampling will eventually replace present methods of 'total' dust sampling using in-line, openface or other types of filter holder. Contrary to popular belief, so-called 'total' dust samplers such as the 37-mm filter cassette do not actually measure true total dust. They are unsuitable for most industrial hygiene monitoring of airborne particles larger than a few micrometers because they not only have high variability but also, importantly, underestimate the contribution of large particles. Even for a fixed sampling flowrate, the extent of this bias is sensitive to windspeed and direction. Implementation of IPM sampling will require a reassessment of ACGIH's *threshold limit values* (TLVs) for aerosols and the continued development and testing of suitable sampling devices.

As explained in the preceding chapters, the bases for criteria for *thoracic particulate matter* (TPM) and *respirable particulate matter* (RPM) are reasonably well established, but the basis for IPM criterion is less well established. Studies during the 1980s have shown that particle inhalation efficiency does not reach zero even for very large particle sizes when data are averaged over all orientations relative to wind direction and over the range of commonly encountered wind velocities (Vincent and Armbruster, 1981; Vincent and Mark, 1981 and 1982; Armbruster and Breuer, 1982; Armbruster *et al.*, 1983). Rather, inhalation efficiency is approximately constant at about 50% for particles larger than 40 µm. As explained elsewhere in this book, the ACGIH Air Sampling Procedures Committee recommended a quantitative criterion for IPM sampling based on these data as early as 1985 (Phalen, 1985), and this was in turn adopted by the International Standards Organisation (ISO, 1992) and the Comité Européen de Normalisation (CEN, 1992). By contrast, earlier ISO recommendations (ISO, 1981 and 1983) were based on the extrapolation of data for particle inhalation efficiency for particles just in the range 0–40 µm, giving an inhalation efficiency curve that decreased to zero at $d_{ae} = 185$ µm.

Sampling to characterize potential biological effects in the gas exchange region of the respiratory tract (respirable aerosol sampling) or the tracheobronchial airway region (tracheobronchial aerosol sampling) is a process that requires initial removal of non-inhalable particles followed by separation into regional fractions. Particles that are not inhaled cannot deposit in the gas exchange region or any other region of the respiratory system. Thus, in principle, all thoracic and respirable sampling requires initial size-selective sampling to eliminate non-inhalable particles. Furthermore, there is an important category of aerosols for which the condition for potential injury is inhalation and deposition anywhere in the body. This includes aerosols of certain metals, radioactive materials, and pesticides. Also included are particles that cause health effects at their site of deposition in the head. Examples include corrosive mists, nickel, chromium aerosols, and wood dust. Thus, inhalable aerosol sampling is fundamental to all aerosol sampling for health risk evaluation.

This chapter includes descriptions and performance of personal and area samplers for inhalable particle sampling. Personal sampling is used to evaluate a worker's exposure to aerosols by measuring the exposure concentration in the worker's breathing zone. The breathing zone is defined as the region within 0.3 m (about 1 ft) of the mouth. Personal sampling is necessary when local particulate sources cause large spatial variation in concentration. Light-weight battery-powered pumps worn on the belt and plastic filter holders clipped to the worker's lapel are used. By contrast, area or static samplers sample from a fixed location, usually at breathing zone height on a stand or tripod. Because the performance of an inhalable sampler depends on the nature of air currents in its vicinity, which in turn are influenced by the presence of any large bluff body (e.g., the body of the wearer), personal samplers will generally not perform well as area samplers and vice versa.

6.2 Inhalable Samplers

Inhalable aerosol sampling can be organized into four categories: 1) scientific studies of the sampling characteristics of thin-wall probes and blunt samplers in still and moving air; 2) standard method approaches, in which the conditions of sampling (e.g., inlet velocity and tube diameter) are specified; 3) standard sampler approaches in which standard hardware is used; and 4) sampling with separation according to particle size so as to match a defined inhalable efficiency curve. The latter is, of course, the most meaningful from the standpoint of evaluating health risk and the most likely to yield exposure data that correlate with disease. In all of these situations, the aerodynamic particle diameter (d_{ae}) is the primary variable that controls sampling efficiency. For the purpose of this chapter, sampler efficiency is expressed in two ways. One is *sampling efficiency*, the ratio of sampler-measured concentration to that in the undisturbed air. The other is *relative sampling efficiency*, the ratio of sampler-based concentration to that based on mannequin inhalation. For the latter a value of 1.0 represents perfect IPM sampling.

6.3 Background to Aerosol Sampling Theory

The simplest and most studied type of sampler is the thin-wall cylindrical probe sampler. Its performance provides insight into the physics of aerosol sampling and serves as a basis for understanding the performance of other types of samplers, including blunt samplers. Sampling in calm air has been evaluated by several investigators (Davies, 1968; Yoshida *et al.*, 1978; Ter Kuile, 1979; Agarwal and Liu, 1980), and summarized by Vincent (1989). These studies deal with the effect of sedimentation and particle inertia on sampling losses.

Numerical particle trajectory analysis has shown the earlier semi-empirical recommendations of Davies to be overly stringent (Agarwal and Liu, 1980).

Probably the most useful model for calm air sampling is the equation of Yoshida *et al.* (1978), based on numerical particle trajectory analysis, that gives the minimum thin-wall probe diameter, D_{min}, necessary to achieve 90% sampling efficiency. Thus

$$D_{min} > 99([v_s]^{2.43})/(g\,U_s^{0.43}) \qquad \textbf{6.1}$$

where v_s is the particle terminal settling velocity [cm/s], U_s is the probe inlet velocity [cm/s] and g is the acceleration due to gravity [cm/s^2]. Equation 6.1 suggests that it is relatively easy to sample 20 µm diameter particles with high efficiency, but very difficult to sample 100 µm diameter particles. Because v_s is proportional to d^2, the minimum probe diameter for representative sampling is roughly proportional to particle diameter raised to the fifth power.

An important assumption underlying calm air sampling is that the air is moving sufficiently slowly that the still-air model holds. Ogden (1983) examined this question and determined that the windspeed should not exceed U_{max}, where

$$U_{max} \leq (0.11\ (Q\tau^2)^{1/3} \qquad \textbf{6.2}$$

where Q is the sampling flowrate [cm^3/s]; τ is the particle relaxation time [s] which, neglecting the slip correction, is given by $\rho_0 d_{ae}^2/18\eta$. Here, ρ_0 is the density of water [1 g/cm^3] and η is the viscosity of the air. This equation suggests that the windspeed must be less than an imperceptible 10 cm/s [20 fpm] for calm air sampling theory to apply to the sampling of 100 µm particles, a size considered to be towards the top end of what can be inhaled. But because of their rapid settling velocity (about 25 cm/s), 100 µm-diameter particles are found in very low concentrations in such calm air. Generally, a vigorous wind or mechanical action is necessary for worker exposure to 100 µm-diameter particles. An exception is where a worker (or sampler) is located directly below the point of generation.

So far the discussion has focused on thin-walled cylindrical probes. The situation is more complicated for the case of blunt samplers sampling in calm air. But Vincent, in his summary of studies of blunt samplers (Vincent, 1989), concludes that, for calm air, the effect of probe bluntness appears to be relatively small.

For thin-walled probes in moving air, the situation becomes different. Now it is the wind that brings particles into the vicinity of the sampler, and not gravitational settling as in the case of pure calm air sampling. For moving air where wind velocities are steady and are greater than given by Equation 6.2, and where particle settling velocities are small compared to windspeed, accurate samples of large particles can be obtained by what is known as "*isokinetic sampling*." This applies to a thin-walled probe that is aligned with the gas streamlines and has an entering air velocity that matches the approaching wind velocity. When these conditions are met, the air-flow streamlines in the vicinity of the entry to the sampler are not distorted, and sampling efficiency is 100% for all particle sizes. Various theoretical and empirical studies have evaluated the effect of misalignment and velocity mismatch as a function of particle size, as reported for example by Durham and Lundgren (1980), Tufto and Willeke (1982), and Vincent *et al.* (1986).

Blunt samplers operating in a wind present a still more complicated situation, and unlike the thin-wall probe situation, there is no unique sampling velocity that permits sampling with 100% efficiency for all particle sizes in a given wind. Results from several investigators have been summarized by Ogden (1983) and by Vincent (1989). For a blunt sampler facing the wind, there is a large scale divergence of approaching streamlines in the vicinity of the sampler and a local convergence of streamlines at the inlet (Ingham, 1981; Vincent *et al.*, 1982; Vincent, 1984). Ogden and Birkett (1977) found that if inlet size is small compared to overall sampler dimensions, the intake flowrate, rather than velocity, is the important parameter to characterize sampling efficiency.

When a blunt sampler has its inlet axis at an angle of 90° from the wind direction, sampling efficiency is less than for face-on and decreases with increasing windspeed and particle size. This is similar to the situation of a blunt sampler with an off-center orifice where the deflected air flow is across the orifice. Sampling efficiency for side-on or off-center orifice blunt samplers is a weaker function of the ratio of wind to sampling velocity than for samplers facing directly into the wind. Recent work by Sreenath *et al.* (1996) has shown that tendency holds also for angles greater than 90°.

One problem with blunt samplers is that known as "*particle blowoff*" which leads to the subsequent sampling of solid particles originally deposited in the immediate vicinity of the inlet. This is less likely to happen for liquid particles. The phenomenon gives rise to differences in collection efficiency (i.e., oversampling) for different types of particles having the same aerodynamic diameter (Vincent and Gibson, 1981; Mark *et al.*, 1982).

Passive aerosol samplers, sometimes referred to as "*dust-fall buckets*," have been used for sampling large particles (Vincent, 1989). Such samplers capture particles by allowing them to settle onto a protected horizontal surface or be projected onto a stationary surface by the action of the external wind alone. They are unsatisfactory for health-related aerosol sampling because they measure the flux of particles onto the collection surface and particulate mass collected in this way cannot be related easily to airborne concentration on which health standards are based.

6.4 Standard Sampling Methods

Some early attempts were made to define "total dust" in terms of what is collected by samplers with specified inlet velocities, inlet diameters, or flowrates. In the early 1970s, for example, Germany established a standard sampling criterion for workplace dust measurements based on such an approach. They defined total airborne particulate matter as that collected by a sampler with an inlet velocity of 1.25 ± 0.12 m/s (MAK, 1981). This was implemented in Europe by the use of field monitor filter holders having a 4-mm diameter inlet and a sampling flowrate of 0.9 L/min. An entry velocity of 1.25 m/s with an orifice diameter of 4–10 mm was considered as a European standard method for sampling lead in the workplace. Another European standards group proposed that samplers should have an inlet velocity of 1.1 to 3 m/s and a flowrate of 0.5 to 4 L/min in order to representatively collect total airborne particulate matter (Ogden, 1983). This corresponds to an inlet diameter range of 2–9 mm.

None of the standard methods referred to attempted to match inhalable efficiency curves. Nor, indeed, was there any basis even to assume that they correctly sampled total airborne particulate matter, except perhaps in calm air conditions. This is unsatisfactory because they are referred to as "total dust samplers," even though they have largely undefined particle-size sampling efficiency characteristics, especially at angles to the wind and averaged over all orientations to the wind. Ogden (1983) estimated that such inlet-velocity criteria would provide greater than 90% sampling efficiency for thin-walled samplers collecting particles smaller than d_{ae} = 36 to 45 μm in calm air. For blunt samplers he estimated a greater than 90% sampling efficiency for particles smaller than d_{ae} = 18–22 μm. These predictions were based on the assumption of calm air, which, for these conditions, corresponds to a wind velocity of less than 0.12 m/s (24 fpm).

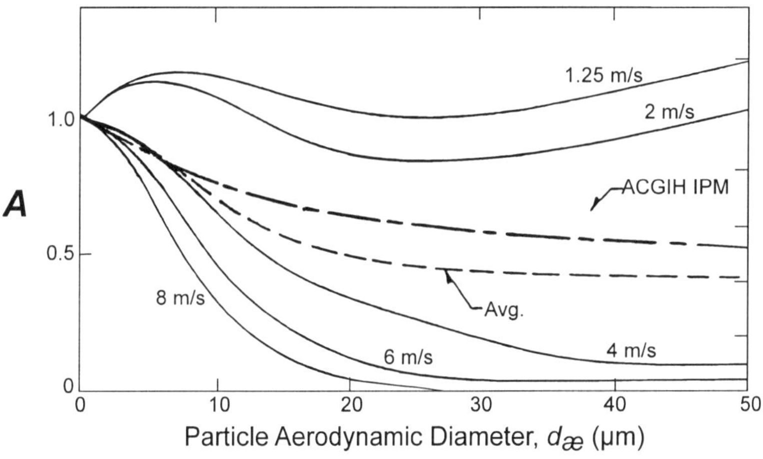

FIGURE 6.1. Sampling perfomance of the Gravicon VC 25G sampler for five wind velocities (Armbruster *et al.*, 1983).

6.5 Standard samplers

"Standard samplers" are samplers of a particular geometric design that have been evaluated in terms of their inhalable sampling performance. Armbruster *et al.* (1983) investigated the sampling characteristics of two sampling instruments that had originally been designed to meet the European inlet velocity criterion cited above, the Gravicon VC25G and the GS050/3. The VC25G samples at 400 L/min and the GS050/3 at 50 L/min. Both have annular (omnidirectional) horizontal inlet slots. The GS050/3 draws air upward through its inlet slot which results in losses by elutriation in the inlet section. The VC25G shows a marked change in sampling efficiency with wind velocity as shown in Figure 6.1. For wind velocities less than 2 m/s (400 fpm), the sampling efficiency versus aerodynamic diameter curve lies significantly above the ACGIH-IPM curve. For wind velocities greater than 4 m/s (800 fpm), the collection efficiency lies well below this curve. Although the curve for **average** efficiency for windspeeds over the full range 1–8 m/s (200–1600 fpm) lies close to the IPM curve, this is not representative of most indoor workplaces, which have windspeeds in the range of 0.1–1.0 m/s (20–200 fpm) except for mines and other special environments. Thus, the VC25G sampler approximately follows the IPM curve only for a narrow range of wind velocities, somewhere in the range 2–4 m/s (400–800 fpm). The sampling efficiency curve averaged for wind velocity for the GS050/3 lies even further from the IPM curve. So neither of these instruments can be considered a satisfactory sampler for IPM.

The sampler described in 1978 by Ogden and Birkett, known as the "Orb," was the first attempt to develop a sampler for inhalable aerosol. It consisted of a sphere 59 mm in diameter with 32 holes, each 2.5 mm in diameter, positioned at 20° latitude and a 6-mm lip positioned at 33° latitude. Sampling flowrate was 2 L/min and a 47-mm diameter filter is mounted at the equator inside the sphere. Its sampling performance was found to satisfactorily match human data for winds of 1 to 3 m/s (200 to 600 fpm), but only for d_{ae} up to about 13 μm. Although this instrument was never used in a practical industrial hygiene setting, and was never commercially available, it is of historical interest because it provided the starting point for subsequent developments.

In 1986, researchers at the Institute of Occupational Medicine, Edinburgh, Scotland proposed a static (or area) sampler that quite closely matched the ACGIH-IPM criterion over the full range of d_{ae} up to about 100 μm (Mark *et al.*, 1985). The sampling head of their device is a vertical axis cylinder about 5 cm in diameter and 6 cm high. A horizontal-axis, oval-shaped inlet slot (about 3 mm high and 16 mm wide) is located midway up the side of the cylinder. The device sampled at 3 L/min through a 37-mm filter mounted in a weighable cassette inside the cylinder. The sampling head was mounted on a larger vertical axis cylinder about 15 cm in diameter and about 18 cm high which housed batteries, pump, and flow control. The sampling head rotated continuously at about 2 rpm. The whole unit weighed 2.5 kg. Performance results are shown in Figure 6.2 for windspeeds in the range 1–3 m/s. Overall, the agreement with the ACGIH-IPM curve is quite good. At a windspeed of 1 m/s, sampling efficiency is within ±10% of the criterion for 0 to 50 μm particles and very close to it for 50 to 100 μm particles. At 3 m/s, the device undersamples in the range of 10 to 50 μm but is within ±10% for other sizes. When data for 1 and 3 m/s windspeeds are averaged, the sampling efficiency shows only a slight departure from the criterion in the range from 30 to 55 μm. This device was commercially available briefly in the late 1980s and early 1990s, but was not widely used and is no longer manufactured. However, higher flowrate prototype versions, up to 70 L/min, have been developed for sampling ambient aerosol (Mark *et al.*, 1995).

6.6 37-MM PLASTIC FILTER CASSETTES

37-mm plastic sampling cassettes have been widely used for many years by industrial hygienists in North America and many other countries. So they have become the de facto basis of many aerosol standards, including ACGIH's TLVs. Fairchild *et al.* (1980) and McCawley *et al.* (1983) evaluated the sampling efficiency of isolated open-face and closed-face 37-mm plastic cassettes used as area samplers at various wind velocities (0–2 m/s) and orientations relative to the sampler axis. It was found that sampling efficiency is strongly biased in favor of particles larger than a few micrometers. When used as an isolated sampler, the cassette seriously oversamples relative to the IPM curve for particles larger than d_{ae} = 10 μm for windspeeds in the range from 0 to 2 m/s (0 to 400 fpm). This suggests that they are of little value as area samplers except to provide a conservative estimate of IPM concentration.

Mark and Vincent (1986) and Chung *et al.* (1987) investigated the orientation-averaged sampling efficiency for three types of filter cassettes similar to the 37-mm cassettes, mounted on a full-torso mannequin. Each was mounted so that its inlet faced forward (axis horizontal) at all times. The open-faced sampler had an inlet diameter of 21 mm; the single-hole sampler, 4-mm; and the seven-hole sampler, seven 4-mm holes. Vincent and Mark's results were originally presented as ratios of concentration based on mass collected by the sampler to that inhaled by the mannequin "breathing" at a minute volume of 20 L and a frequency of 20 breaths per minute. Their original published data have since been replotted in terms of sampling efficiency as a function of d_{ae}, and they are shown in this form in Figure

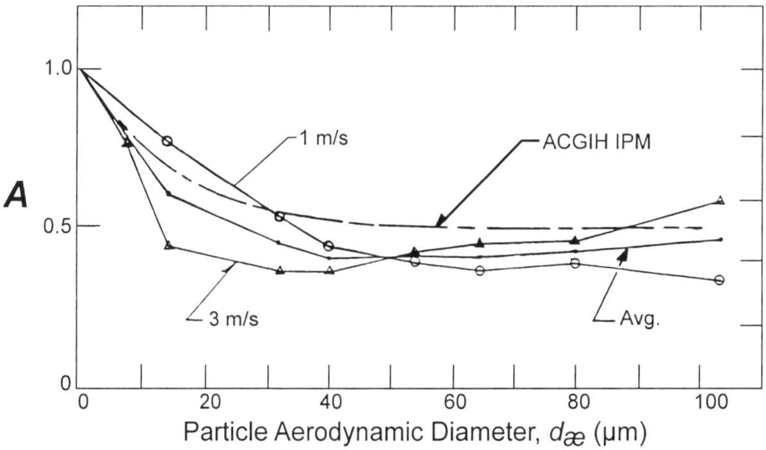

FIGURE 6.2. Orientation-averaged performance of the IOM static sampler at two windspeeds (Mark *et al.*, 1985).

6.3. This figure shows that all three samplers undersampled significantly with respect to the IPM curve.

Similar data for 37-mm plastic cassettes mounted on a plate facing the wind (simulating the wearing of the sampler on the torso of a

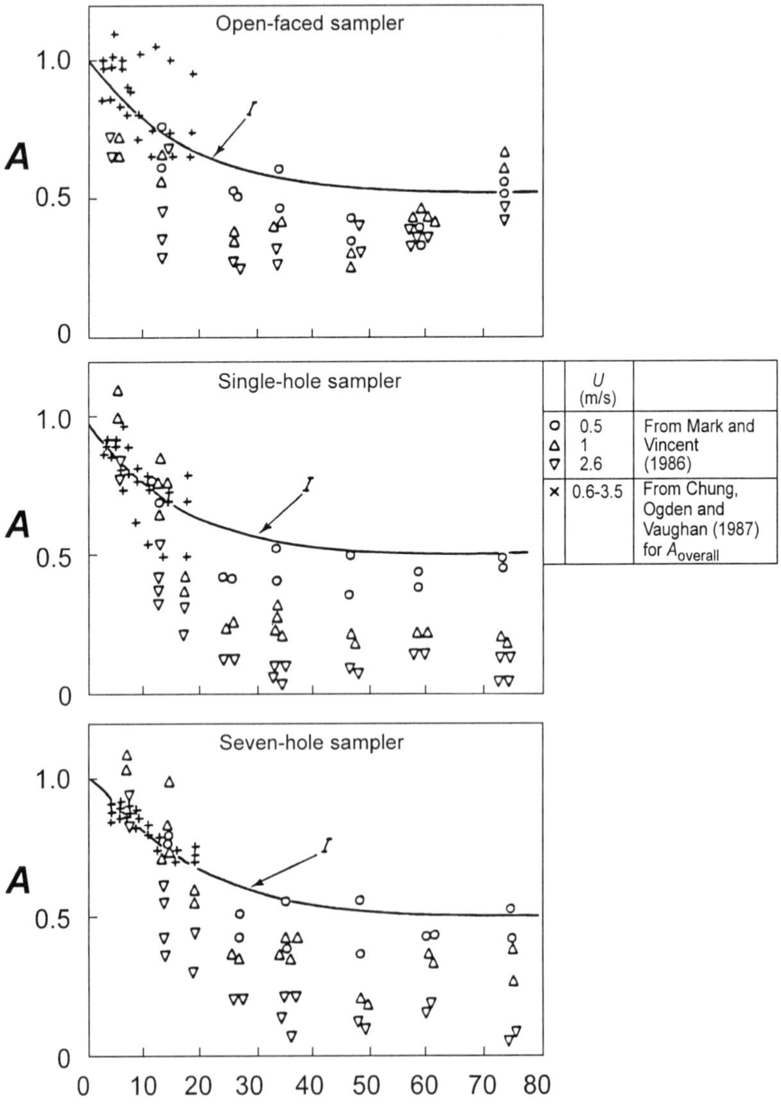

FIGURE 6.3. Sampling performance of three personal sampling filter cassettes (Vincent, 1989).

worker) were presented in 1986 by Buchan *et al.* (1986). They evaluated sampling efficiency for open-face and closed-face 37-mm cassettes facing a wind of 1.0 m/s. The cassettes were either hanging free with their inlet axes aligned vertically downward, or attached to the plate with their inlet axes at 46° below the horizontal. In all cases the downward facing sampler showed lower sampling efficiency than the 46° angled sampler. For inline cassettes, the ratio of down to 46° sampling efficiency ranged from 0.73 for 2.4 μm to 0.23 for 24 μm particles. This underscores the sensitivity of sampler performance to the orientation of the inlet.

A recent report of a collaborative European study by Kenny *et al.* (1997) has confirmed this tendency, as well as those of the samplers studied earlier by Mark and Vincent. Figure 6.4 shows the results for the 37-mm sampler and indicates that the instrument undersamples significantly for the larger particles.

Fairchild *et al.* (1980) and Doemeny and Smith (1981) had earlier recommended the use of the closed-face 37-mm plastic cassette for total dust (total airborne particulate matter) sampling because it had exhibited sampling efficiency closest to unity in their tests. However, based on the weight of the more recent information, this is no longer considered to be true. This presents a dilemma for industrial hygienists who are continuing to use the 37-mm sampler in the field. The question is: how do the results obtained with the 37-mm sampler relate to IPM? Based on extensive side-by-side, 'total' and inhalable (see below) inter-sampler comparisons in a wide range of industrial settings, Werner *et al.* (1996) have recently proposed a range of conversion factors for estimating inhalable aerosol concentration based on measured 'total' aerosol concentration using the closed-face 37-mm cassette (see Table 6.1). The coarser aerosol types have the largest conversion factor.

6.7 Inhalable aerosol samplers

In 1986, Mark and Vincent proposed a personal sampler specifically designed to match the ACGIH-IPM curve. The device has a cylindrical body, 37-mm in diameter and 27-mm long, with a 15-mm diameter inlet which faces forward. It uses a 25-mm filter contained in a light cassette having a 15-mm diameter inlet tube that protrudes 1.5 mm beyond the body of the sampler. The filter and cassette are weighed together. Tests with the original prototypes were conducted in a wind tunnel with the sampler mounted on the torso of a mannequin and with a sampling flowrate of 2 L/min. The orientation-averaged sampling efficiency for windspeeds in the range from 0.5 to 2.6 m/s is shown in Figure 6.5. Overall, it is seen that agreement with the IPM curve is very good. The first commercial versions of the instrument

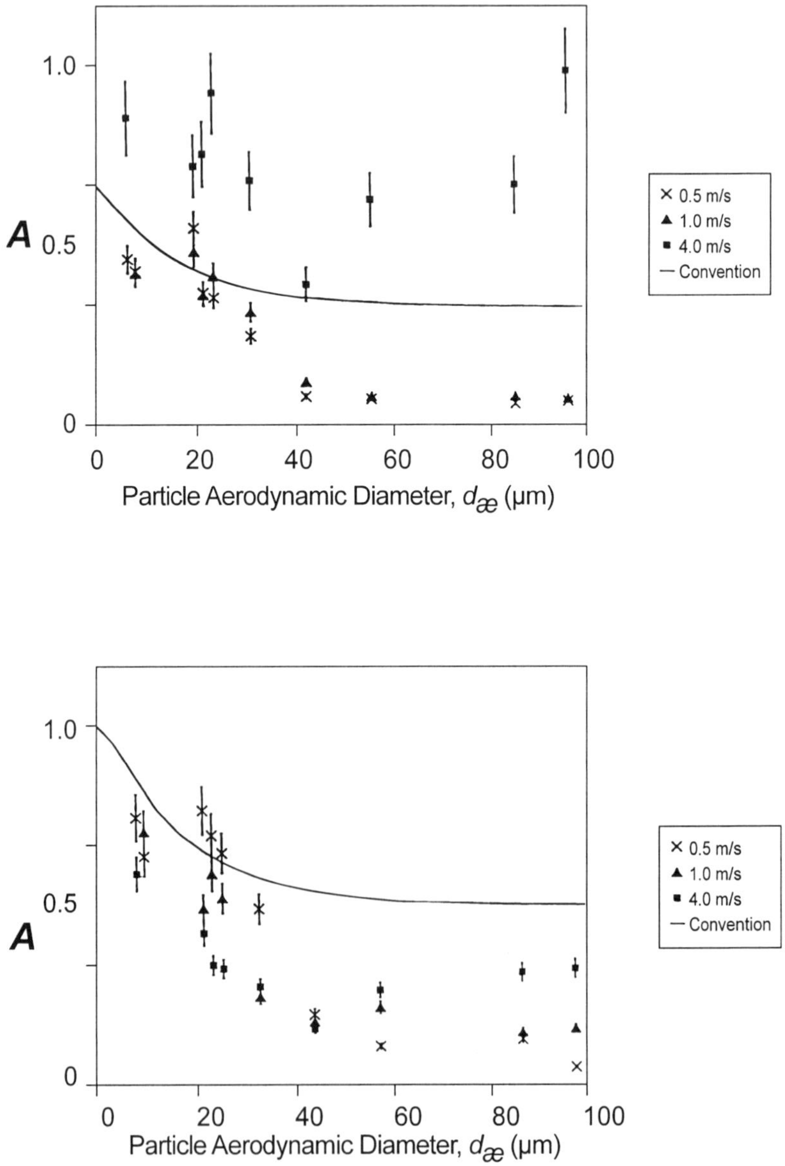

FIGURE 6.4. Sampling efficiency for open-faced and closed-face 37-mm cassettes (Kenny *et al.*, 1997).

TABLE 6.1. Suggested working converson factors for estimating inhalable concentration based on 37-mm cassette measured concentration *

Aerosol/Activity	Suggested Conversion Factor
Dust	
Mining, ore and rock handling	2.5
Handling bulk aggregates	
Textiles, flour, and grain	
Mist	
Oil mist, machining fluid, paint spray	2.0
Electroplating	
Hot Processes	
Foundries, smelting and refining	1.5
Welding, smokes, and fumes	1.0

*Source: Werner *et al.* (1996)

were made from aluminum (including the cassette). But these have been discontinued. More recently, an electrically-conductive plastic version of this sampler has become available from SKC, Inc. (Eighty Four, PA). This version was tested as part of the collaborative European study, and these new results are shown in Figure 6.6. For the lower windspeeds, they are reasonably consistent with the earlier data, although they show more clearly the slight oversampling (with respect to IPM) at 0.5 m/s and slight undersampling at 1 m/s. At the higher windspeed of 4 m/s, there is significant undersampling.

In a recent study, Kenny *et al.* (1999) have reported new experiments — at two separate laboratories — with the IOM sampler at still lower windspeeds, stimulated by field evidence that actual windspeeds in many workplaces tend to be even lower than the lowest ones used in the preceding reported studies. They were carried out with the mannequin–sampler combination in a test chamber in which the mannequin was rotated slowly in order to simulate the low windspeeds of interest. Summary results are shown in Figure 6.7. These are averages for the two laboratories, where inspection of the raw data shows good general agreement between the two laboratories. Figure 6.7, like Figure 6.6, shows sampling efficiency at very low windspeed being significantly above the IPM curve. While this indicates a clear breakdown of the agreement between the performance of the IOM sampler and the formal IPM criterion, it is noted that the results do in fact lie very close to the corresponding measured inhalability at the same low windspeed (see also Chapter 3). So these results show that the IOM sampler continues to measure the fraction that a worker would actually inhale, even at very low windspeeds.

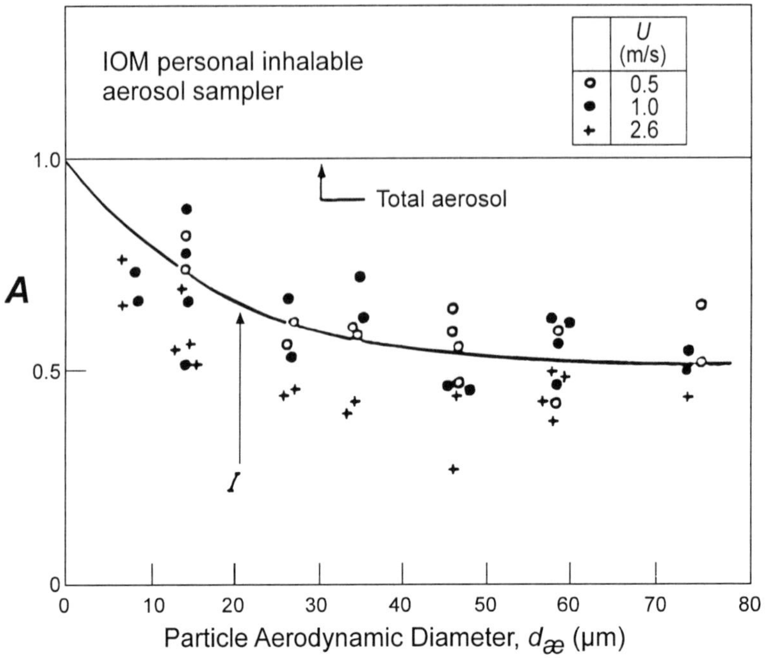

FIGURE 6.5. Earlier results for the sampling efficiency of the IOM personal inhalable particulate matter sampler (Vincent, 1989).

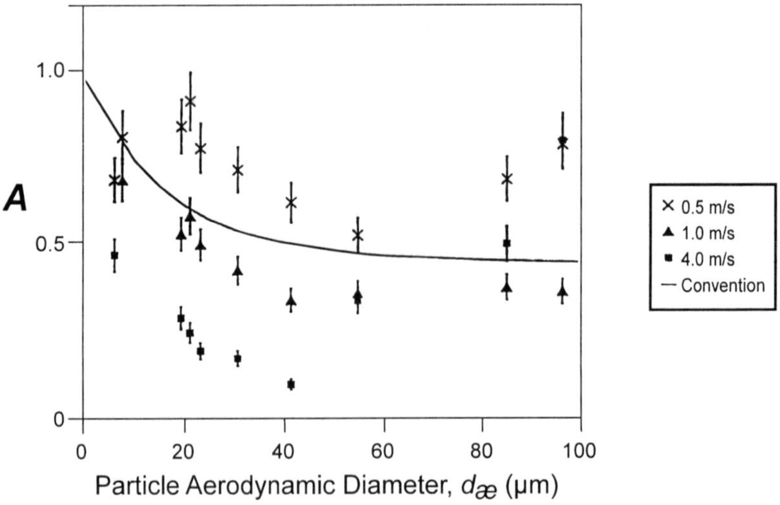

FIGURE 6.6. Recent results for the sampling efficiency of the IOM sampler (Kenny *et al.*, 1997).

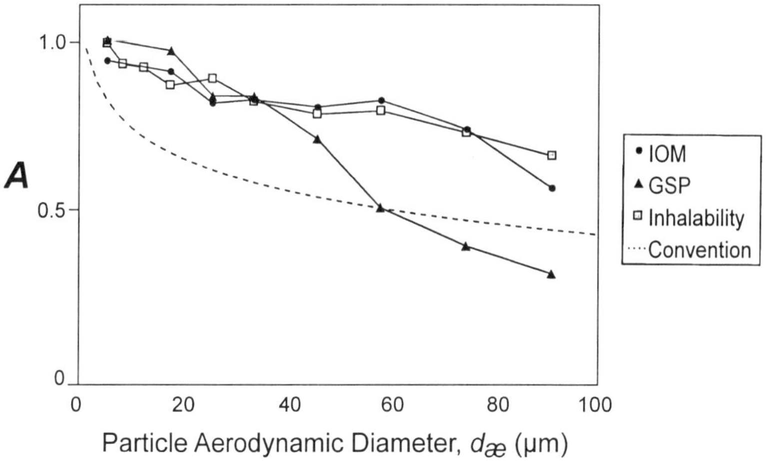

FIGURE 6.7. Sampling efficiency for the IOM and GSP samplers at very low windspeeds (Kenny *et al.*, 1999).

Smith *et al.* (1998) described problems with the large moisture absorption of the plastic filter cartridge in one version of the IOM sampler as supplied by SKC Inc. and noted that the stainless steel cartridge is much better in this regard.

Several other samplers have been proposed whose performances come close to matching the IPM curve. Vincent and Mark (1990) tested the personal aerosol spectrometer (PERSPEC) sampler which was developed in Italy (Prodi *et al.*, 1986 and 1992). This instrument seeks to measure the inhalable, thoracic, and respirable fractions simultaneously. It samples at 2 L/min and collects particles on a 47-mm filter. Larger particles are collected near the center of the filter. By following a prescribed method of cutting the filter, the TPM and RPM fractions can be determined. As shown in Figure 6.8, the aspiration efficiency of the instrument is in reasonable agreement with the IPM curve. More recently, Dunkhorst *et al.* (1995) have described a personal and area inhalable sampler — the “Respicon” — for similarly collecting the three fractions simultaneously. This device has an inhalable inlet followed by two virtual impactors in series to separate the aerosol stream into the thoracic and respirable fractions. It is 10 cm high, weighs 390 g and samples at 3 L/min. In addition to filter collection at each stage, one version has photometric sensors to provide continuous electronic output to a data logger. The device gives a reasonable match to the inhalability criterion as an area sampler in winds up to 2.5 m/s. It is commercially available from TSI,

Inc., (St. Paul, MN). A prototype modified version of the IOM personal sampler, also designed to collect the inhalable, thoracic, and respirable fraction simultaneously, has been described by Mark *et al.* (1988). The inlet geometry is the same as the previously described IOM personal sampler. In this instrument, however, the aerosol passes through two porous foam selectors with penetration characteristics matching the TPM and RPM criteria and a final filter. Particles collected on the final filter represent the RPM fraction. Particles collected on the final filter and the second foam selector together represent TPM and all collected particles represent IPM. Another personal inhalable-inlet sampler is described by Gibson *et al.* (1987). It uses an IOM inlet, 15-mm ID at 2 L/min, to aspirate the inhalable fraction which is then size classified by a small cascade impactor. The TPM and RPM fractions can be calculated from the particle size distribution by mathematical inversion of the raw data provided by the cascade impactor.

In addition to studying the IOM sampler, the European research reported by Kenny *et al.* (1997) also obtained performance data for several other instruments and related them to the IPM curve. These included the German GSP (see Figure 6.9a) and the Dutch PAS-6 (see Figure 6.9b), both of which were not studied during the earlier Vincent and Mark research. The results show that both of these samplers provide a fair match with the IPM curve at the lower windspeeds. The research also included further tests of the Italian PERSPEC, as well as the French CIP10 (Courbon *et al.*, 1988), by Vincent and Mark. While the PERSPEC again exhibited a fair match with the IPM curve at low windspeed, the CIP10 was seen to oversample significantly.

Figure 6.7 also shows some results for the GSP sampler at very low windspeed (approaching calm air), and it is seen here that its sampling efficiency falls below that of the IOM for larger particle sizes. While this trend brings the GSP results closer to the conventional IPM curve, it actually means that this sampler oversamples with respect to true inhalability at larger particle sizes (for the reason suggested above, see also Chapter 3).

6.8 OTHER APPROACHES TO IPM SAMPLING

The presence of airborne particles larger than a certain size can be determined by isokinetic sampling, a rotating arm sampler, or with settling plates and microscopic examination. For the latter, however, the use of settling plates is only a screening technique to detect the

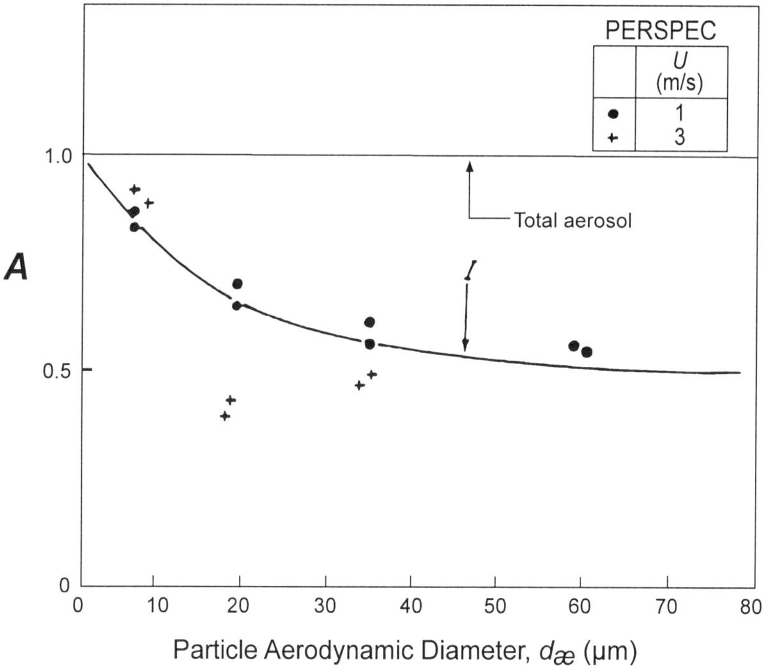

FIGURE 6.8. Sampling efficiency for the Italian PERSPEC aerosol spectrometer (Vincent, 1989).

possible presence of large particles and is not intended to quantitatively measure their concentration. An oil-coated glass slide may be placed horizontally at the breathing zone for at least one hour. The presence of more than one 20 μm particle in each square centimeter indicates a significant contribution of this size particle in a situation where the TLV is 0.001 mg/m^3. Numerically larger TLVs require proportionately greater numbers of particles.

In theory, an IPM sample can be obtained by collecting a (true) total dust sample and passing it through a separator that removes non-inhalable particles, thus permitting the inhalable particles to be collected on a filter. Unfortunately, there are at present no in-line separators that match the inhalable sampling criterion. Alternatively, a true total dust sample can be used to estimate IPM, recognizing that it will overestimate IPM concentration and thus be conservative. When wind velocities exceed 1 m/s and are reasonably constant, isokinetic sampling can be used to obtain a valid total dust sample. Here, the inlet sampling velocity should be at least two times the settling velocity of the largest particles of interest. For example, 100 μm and 200 μm diameter "unit" density particles (i.e., density of 1 gm/cm^3) have a settling velocity of 25 and 71 cm/s (49 and 140 fpm), respectively. A second approach, suitable for low wind velocities, is to use a rotating arm type total mass sampler similar to that described by Hameed *et*

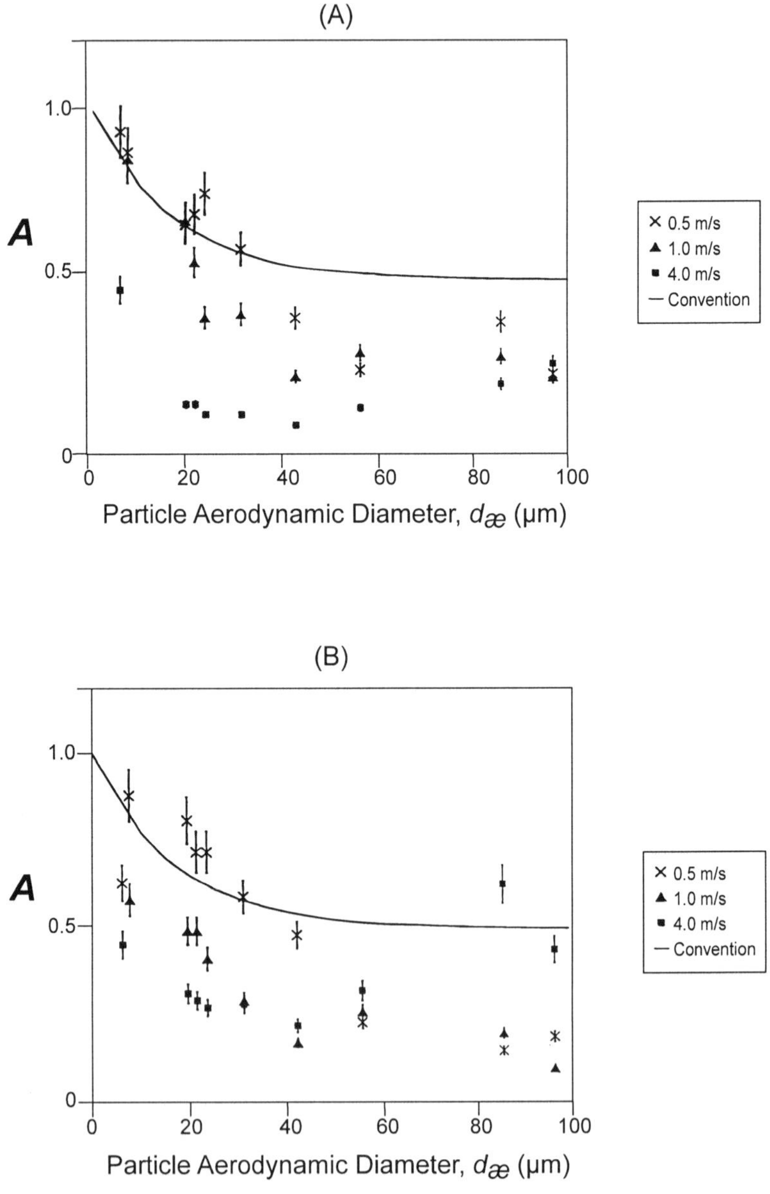

FIGURE 6.9. Sampling efficiency for two personal samplers: a) the German GSP; and b) the Dutch PAS-6 (Kenny *et al.*, 1997).

al. (1983) (although it is not clear that a device of this type can be used in a typical work environment unless surrounded by a protective enclosure).

Another approach to inhalable aerosol sampling is to directly simulate the human inhalation situation. Guidelines for such a sampler might be:

1. The sampling head has a gross dimension that is within 50% of a normal adult head, 18 cm in diameter. A thorax-like support may also be necessary.
2. It has inlet dimensions representative of the human mouth, 1- to 2-cm diameter circular inlets or equivalent area noncircular inlets.
3. It has an inlet axis in the horizontal plane. For omnidirectional samplers, a continuous slit in the horizontal plane of 0.3 to 1 cm high with the same inlet velocity range as provided for circular inlets.
4. It has a sampling flowrate of 43.5 L/min equal to the average inhalation flowrate for the reference worker (Snyder *et al.*, 1975).

It is recognized, however, that these conditions may not be typical, nor do they represent the extreme worst case. But they are representative of exposure conditions likely to be experienced for a full eight-hour shift. In more severe cases, results should be adjusted or interpreted accordingly. Samplers should be mounted at a height so that the inlet is within 30 cm (one foot) of the height of a typical worker's mouth. Samplers may be mounted on the worker or may take the form of stationary area samplers that are continuously rotated, or have omnidirectional inlets. One example of such a personal sampler would be an inlet and filter mounted on or in a worker's hard hat and constructed and operated so that it meets the guidelines described above.

REFERENCES

Agarwal, J.K. and Liu, B.Y.H. (1980), A criterion for accurate aerosol sampling in calm air, *Am. Ind. Hyg. Assoc. J.*, 41, pp. 191-197.

Armbruster, L. and Breuer, H. (1982), Investigations into defining inhalable dust, In: Inhaled Particles V (W.H. Walton, Ed.), Pergamon Press, Oxford, U.K., pp. 21-32.

Armbruster, L.; Breuer, H.; Vincent, J.H. et al. (1983), Definition and measurement of inhalable dust, Aerosols in the Mining and Industrial Work Environment (V.A. Marple and B.Y.H. Liu, Eds.), Ann Arbor Science Publishers, Ann Arbor, MI.

Buchan, R.M., Soderholm, S.C. and Tillery, M.I. (1986), Aerosol sampling efficiency of 37-mm filter cassettes, *Am. Ind. Hyg. Assoc. J.*, 47, pp. 825-831.

Chung, K.Y.K., Ogden, T.L. and Vaughan, N.P. (1987), Wind effects on personal dust samplers, *J. Aerosol Sci.*, 18, pp. 159-174.

Comité Européen de Normalisation (CEN) (1992), Workplace atmospheres: size fraction definitions for measurement of airborne particles in the workplace, CEN Standard EN 481.

Courbon, P., Wrobel, R. and Fabries, J-F. (1988), A new individual respirable dust sampler: the CIP 10, *Ann. Occup. Hyg.*, 32, pp. 129-143.

Davies, C.N. (1968), The entry of aerosols into sampling tubes and heads, *Brit. J. Appl. Phys.*, 25, pp. 921-932.

Doemeny, L.J. and Smith, J.P. (1981), Letter to the Editor, *Am. Ind Hyg. Assoc. J.*, 42, A22.

Dunkhorst, W., Lodding, H. and Koch, W. (1995), A new personal monitor for real-time measurement of the inspirable, the thoracic, and the respirable fraction of airborne dust, *J. Aerosol Sci.*, 26, S107-S108.

Durham, M.D. and Lundgren, D. (1980), Evaluation of aerosol aspiration efficiency as a function of Stokes number, velocity ratio and nozzle angle, *J. Aerosol Sci.*, 11, pp. 179-188.

Fairchild, C.I., Tillery, M.I., Smith, J.P. *et al.* (1980), Collection efficiency of field sampling cassettes, LA8640-MS. Los Alamos Scientific Laboratory.

Gibson, H., Vincent, J.H. and Mark, D. (1987), A personal inspirable aerosol spectrometer for application in occupational hygiene research, *Ann. Occup. Hyg.*, 31, pp. 463-479.

Hameed, R., McMurry, P.H. and Whitby, K.T. (1983), A new rotating coarse particle sampler, *Aerosol. Sci. Tech.*, 2, pp. 69-78.

Ingham, D.B. (1981), The entrance of airborne particles into a blunt sampling head, *J. Aerosol Sci.*, 12, pp. 541-549.

International Standards Organisation Ad Hoc Working Group to Technical Committee TC146 (1981), Size definitions for particle sampling, *Am. Ind. Hyg. Assoc. J.*, 42, A64-A68.

International Standards Organisation (ISO) (1983), Air quality -particle size fraction definitions for health-related sampling, ISO/TR 7708-1983(E).

International Standards Organisation (ISO) (1992), Air quality—particle size fraction definitions for health-related sampling, ISO/TR 7708-1983(E).

Kenny, L.C., Aitken, R.J., Chalmers, C. *et al.* (1997), A collaborative European study of personal inhalable aerosol sampler performance. *Ann. Occup. Hyg.*, 41, pp. 135-153.

Kenny, L.C., Aitken, R.J., Baldwin, P.E.J. *et al.* (1999), The sampling efficiency of personal inhalable aerosol samplers in low air movement environments, *J. Aerosol Sci.*, in press.

MAK (1981), Mitteilung der Senatskammission zur Prufung Gesundheitsschadlicher Arbeitsstoffe, Maximale Arbeitsplatz konzentation, Harald Boldt Verlag, Boppard, Germany.

Mark, D. and Vincent, J.H. (1986), A new personal sampler for airborne total dust in workplaces, *Ann. Occup. Hyg.*, 30, pp. 89-120.

Mark, D., Vincent, J.H. and Witherspoon, W.A. (1982), Particle blowoff: a source of error in blunt dust samplers, *Aerosol Sci. Tech.*, 1, pp. 463-469.

Mark, D. Vincent, J.H., Gibson, H. *et al.* (1985), A new static sampler for airborne total dust in workplaces, *Am. Ind. Hyg. Assoc. J.*, 46, pp. 127-133.

Mark, D., Borzucki, G., Lynch, G. *et al.* (1988), The development of personal sampler for inspirable, thoracic, and respirable aerosol, Presented at the annual conference of The Aerosol Society, Bournemouth, U.K., March 1988.

Mark, D., Upton, S.L., Hall, D.J. *et al.*(1995), Improvements to the design and performance of an ambient inhalable aerosol sampler, AEA Technology Report, AEA-TPD-0299, Harwell, UK.

McCawley, M.A., Burkhart, J., Baron, P.A. *et al.* (1983), Testing of a personal filter cassette with a circumferential orifice, Presented at American Industrial Hygiene Conference, Philadelphia, PA, May 1983.

Ogden, T.L. (1983), Inhalable, inspirable, and total dust, In: *Aerosols in the Mining and Industrial Work Environment* (V.A. Marple and B.Y.H. Liu, Eds.), Ann Arbor Science, Ann Arbor, MI.

Ogden, T.L. and Birkett, J.L. (1977), The human head as a dust sampler, In: *Inhaled Particles IV* (W.H. Walton, Ed.), pp. 93-105, Pergamon Press, Oxford.

Ogden, T.L. and Birkett, J.L. (1978), An inhalable dust sampler for measuring the hazard from total airborne particulate, *Ann. Occup. Hyg.*, 21, pp. 41-50.

Phalen, R.F. (Ed.) (1985), Particle size-selective sampling in the workplace, Report of the ACGIH Air Sampling Procedures Committee, American Conference of Governmental Industrial Hygienists, Cincinnati, OH.

Prodi, V., Belosi, F. and Mularoni, A. (1986), A personal sampler following ISO recommendations on particle size definitions, *J. Aerosol Sci.*, 17, pp. 576-581.

Prodi, V., Sala, F. and Belosi, F. (1992), PERSPEC, personal size separating sampler: operational experience and comparison with other field devices, *Appl. Occup. Environ. Hyg.*, 7, pp. 368-374.

Smith, J.P., Bartley, D.L. and Kennedy, E.R. (1998), Laboratory investigation of the mass stability of sampling cassettes from inhalable aerosol samplers, *Am. Ind. Hyg. Assoc. J.*, 59, pp. 582-585.

Snyder, W.S. (1975), Report of the Task Group on Reference Man, International Commission on Radiological Protection (ICRP), Pergamon Press, Oxford.

Sreenath, A., Ramachandran, G. and Vincent, J.H. (1996), Experimental studies of the aspiration efficiencies of blunt samplers at large angles to the wind, *J. Aerosol Sci.,* 27, pp. S677-S678.

Ter Kuile, W.M. (1979), Dust sampling criteria, *J. Aerosol Sci.*, 10, pp. 241-242.

Tufto, P.A. and Willeke, K. (1982), Dynamic evaluation of aerosol sampling inlets, *Environ. Sci. Technol.*, 16, pp. 607-609.

Vincent, J.H. (1984), A comparison between models for predicting the performance of blunt dust samplers, *Atmos. Environ.*, 18, pp. 1033-1035.

Vincent, J.H. (1989), *Aerosol Sampling: Science and Practice*, Wiley, Chichester, England, U.K.

Vincent, J.H. and Armbruster, L. (1981), On the quantitative definition of the inhalability of airborne dust, *Ann. Occup. Hyg.*, 24, pp. 245-248.

Vincent, J.H. and Gibson, H. (1981), Sampling errors in blunt dust samplers arising from external wall loss effects, *Atmos. Environ.*, 15, pp. 703-712.

Vincent, J.H. and Mark, D. (1981), The basis of dust sampling in occupational hygiene: a critical review, *Ann. Occup., Hyg.*, 24, pp. 375-390.

Vincent, J.H. and Mark, D. (1982), Application of blunt sampler theory to the definition and measurement of inhalable dust, In: *Inhaled Particles V* (W.H. Walton, Ed.), Pergamon Press, Oxford, U.K., pp. 3-19.

Vincent, J.H. and Mark, D. (1990), Entry characteristics of practical workplace aerosol samplers in relation to ISO recommendations, *Ann. Occup. Hyg.*, 34, pp. 249-262.

Vincent, J.H., Hutson, D. and Mark, D. (1982), The nature of air flow near the inlets of blunt dust sampling probes, *Atmos. Env.*, 16, pp. 1243-1249.

Vincent, J.H., Stevens, D.C., Mark, D., Marshall, M., and Smith, T.D. (1986). On the entry characteristics of large-diameter, thin-walled aerosol sampling probes at yaw orientations with respect to the wind. J. Aerosol Sci., 17, pp. 211-224.

Werner, M.A., Spear, T.M. and Vincent, J.H. (1996), Investigation into the impact of introducing workplace aerosol standards based on inhalable fraction, *The Analyst*, 121, pp. 1207-1214.

Yoshida, H., Uragami, M., Masuda, H. *et al.* (1978), Particle sampling efficiency in still air, *Kagaku Kagaku Robunshu*, 4, pp. 123-128.

Chapter 7

SAMPLING FOR THORACIC AEROSOL

Paul Baron[1] **and Walter John**[2]

[1]*Centers for Disease Control and Prevention, National Institute for Occupational Safety and Health, Cincinnati, OH*

[2] *Particle Science, Walnut Creek, CA*

7.1 General Requirements — The Ideal Sampler

There are two possible approaches to sampling for thoracic particulate matter (TPM). The simplest is to use a sampler whose collection efficiency as a function of particle aerodynamic diameter (d_{ae}) satisfies the acceptance criteria. Such a TPM sampler consists of an inlet, a size-fractionating stage, which is sometimes integral with the inlet, and a particle collector, which is usually a filter. The U.S. Environmental Protection Agency (EPA) has defined the PM_{10} sampling convention to have a 50% cut point at $d_{ae} = 10$ µm. The slope of the PM_{10} curve is similar to, though differs slightly from, the TPM curve (see Figure 7.1) (John, 1993). The most significant difference is at the large particle ends of the curves. The bias of the PM_{10} convention relative to the thoracic convention is calculated and presented in Figure 7.2 for a range of lognormal particle size distributions. This figure shows that although the difference is small over a wide range of conditions, it can become significant if the particle size distribution being sampled is rich in coarse particles. It should be noted, however, that some PM_{10} samplers have an efficiency curve with a large-particle tail approximating that of the thoracic curve (see Figure 7.1).

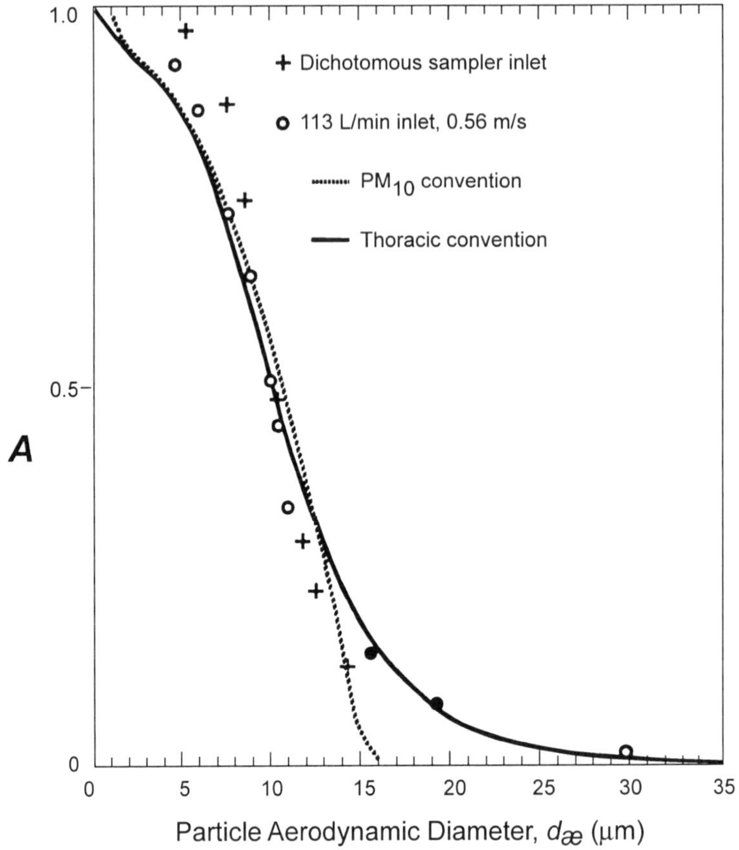

FIGURE 7.1. Thoracic and PM_{10} sampling criteria with data for two PM_{10} samplers — the dichotomous sampler (Wedding, 1982) and the 113 L/min sampler (Wedding, *et al.* 1983)

An alternative approach is to determine the aerodynamic particle size distribution; for example, by sampling with a cascade impactor. The thoracic mass fraction is then calculated from the data and the TPM criteria. This approach provides more detailed information, but it has disadvantages including a large increase in the number of samples to be analyzed, the complexity of the required calculations, and the limited collection capacity of cascade impactors.

The measurement of TPM can be carried out both for personal exposure and for environmental monitoring. For personal monitoring, the inlet of the sampler should comply with the requirements for an inhalable inlet with a pre-classifier selecting the TPM.

The sampling of particles having d_{ae} = 10 μm and somewhat larger for environmental monitoring requires an inlet designed to be independent of wind direction and windspeed (Wedding *et al.*, 1977 and 1980; Liu and Pui, 1981; Wedding, 1982; Tufto and Willeke, 1982). Indoor windspeeds are of the order of 0.1–1 m/s, while mine and outdoor environments have higher windspeeds (Berry and Froude, 1989). The EPA (1987) requires PM_{10} (particulate matter, 10-μm cut point) samplers for ambient air to be tested at 2, 8 and 24 km/hr (0.56 to 6.67 m/s). Testing of a thoracic sampler should include placing it on a mannequin and ascertaining the susceptibility of its inlet to wind direction and velocity biases. The sampling efficiency is measured at several wind velocities between stagnant air and 4 m/s and averaged over 360°. Future standards may select specific wind velocities and orientations for the test protocol. Little such testing for thoracic samplers has so far been carried out.

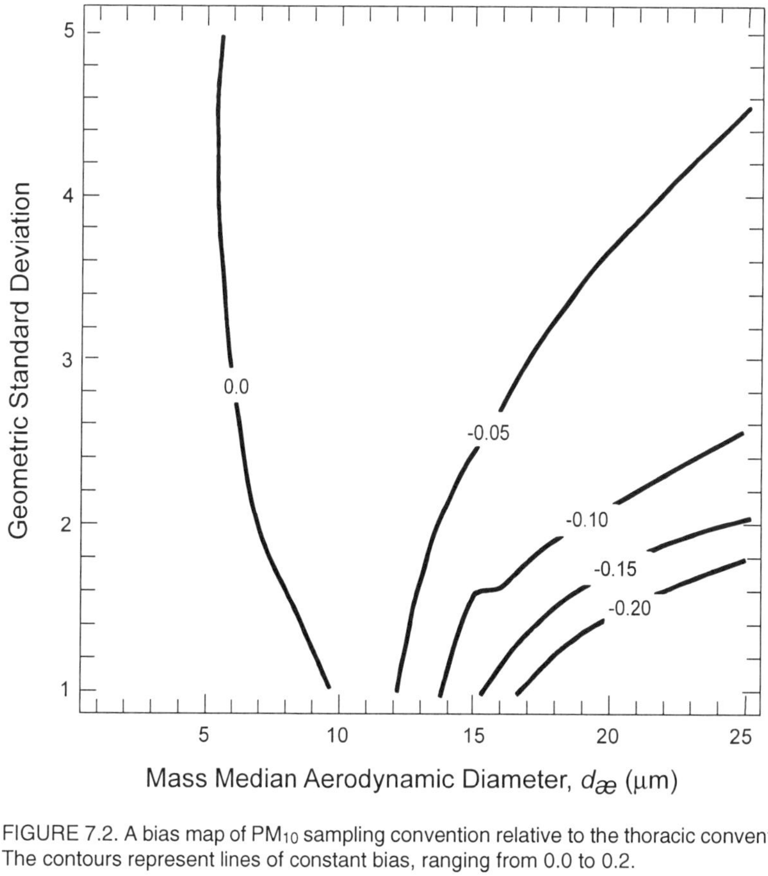

FIGURE 7.2. A bias map of PM_{10} sampling convention relative to the thoracic convention. The contours represent lines of constant bias, ranging from 0.0 to 0.2.

TABLE 7.1. Principal types of filter media for sampling

Type	Notable Characteristics
Cellulose fiber	Initial efficiency low, increasing rapidly with loading, hygroscopic, low ash, high purity
Glass fiber	Low pressure drop, produces sulfate and nitrate artifacts, tolerates high temperatures, high loading capacity
Quartz fiber	Brittle, low artifact formation
Plastic fiber	Lower pressure drop than membranes, poor mechanical strength
Membrane	Available in cellulose nitrate, cellulose triacetate, polyvinyl chloride, Nylon, polypropylene, polyimide, polysulfone, Teflon and silver. High filtration efficiency, particles retained near surface, useful for microscopy, X-ray fluorescence analysis, clog easily
PVC membrane	Low hygroscopicity
Teflon membrane	Low mechanical strength, high purity, low artifact formation
Silver membrane	Useful for X-ray diffraction analysis, organic analysis
Nuclepore	Unique straight-through pores allow penetration of particles smaller than pores, solid particles can bounce and penetrate. Polycarbonate material is strong, non-hygroscopic, high purity. Useful for microscopy, X-ray fluorescence

Note: For more information on filter media, see Lippmann (1995).

The TPM criterion requires aerodynamic separation or sizing of the particles. In practice this is frequently accomplished by inertial impaction, which implies the need for precautionary measures against excessive penetration due to particle bounce and reentrainment (Rao and Whitby, 1977; Wesolowski *et al.*, 1977). Also, deagglomeration can lead to excessive penetration (John *et al.*, 1991).

The choice of filter media for particle collection depends on the physical and chemical properties of the particles to be sampled, the sampler, and the analysis to be performed. Important filter characteristics (see Table 7.1) are the collection efficiency (John and Reischl, 1978; Lippmann, 1995), the pressure drop, mechanical strength, hygroscopicity, chemical purity and possible artifact production (gas-to particle conversion). The filter holder must seal securely against air leakage around the filter or to the outside. The sampler should have a flow controller to stabilize the flowrate as the pressure drop increases with filter loading or the environmental temperature changes.

The current state of knowledge of fluid and aerosol mechanics permits the basic design of TPM samplers from first principles. However, some important details involve empiricism. So the performance of the prototype sampler must be verified in the laboratory. Inlet testing involves the use of aerosol generators in wind tunnels. Subsequently, the samplers are tested in the field to investigate real-world problems. Obviously, samplers vary in the degree to which they approach ideal performance. So the user must choose a sampler with acceptable performance which also satisfies the particular requirements of the application.

7.2 THORACIC SAMPLING DEVICES

The selection of a TPM sampler is made, first, by considering whether personal or environmental sampling is required and, second, by the flowrate. The first commercial TPM personal sampler was the so-called "personal environmental monitor" (or PEM) (Marple, 1989; Buckley *et al.*, 1991) which was aimed at personal exposure based on the PM_{10} environmental standard. Additional personal samplers have only recently been developed subsequent to the adoption of the sampling conventions. Although there has therefore been relatively little evaluation of these devices, it is an area of active research. Because of the limited availability of TPM samplers, the discussion here and Table 7.2 include samplers that vary in the degree with which they satisfy the TPM criteria. It also includes samplers which at present exist only as prototypes and so are not yet commercially available. The user must make an informed choice of sampler depending on the application and actual availability.

Environmental TPM samplers can be classified into low volume ($Q < 20$ L/min), medium volume (20 L/min $< Q <$ 150 L/min) and high volume ($Q >$ 150 L/min) samplers (see Table 7.2). Samplers are currently available in each of these ranges in part as a result of the sampler development program of the EPA in support of its particulate standard for ambient air (EPA, 1987; John, 1993).

Several personal samplers were designed for the collection of the thoracic particulate mass fraction. A personal environmental monitor developed by Marple has a single, sharp impactor cut at d_{ae} = 10 µm (Marple, 1989; Buckley *et al.*, 1991). The PEM (see Figure 7.3) collects particles larger than 10 µm on an oiled substrate, and the thoracic fraction is collected on a downstream filter. The CIP10 sampler used in France was modified to include the measurement of thoracic as well as respirable dust (Fabries *et al.*, 1989). This sampler uses the inertia of air within a rotating head to draw air through porous foam that collects the thoracic fraction. Thus, the air moving device is integral to the particle size-selector of the CIP10 sampler. Particles smaller

TABLE 7.2. Thoracic particulate fraction samplers

Flowrate	Sampler
A. Personal	
1.6 L/min (not verified)	Kenny-Gussman (Kenny and Gussmann, 1997)
2 L/min	PEM (Marple, 1989)
2 L/min	Vincent foam sampler (Vincent *et al.*, 1993)
5.2 L/min	CIP 10 (Fabries *et al.*, 1987)
2 L/min	Cascade impactors (Gibson *et al.*, 1987; Rubow *et al.*, 1987; Hering, 1995)
B. Low	
6 L/min	NBS portable ambient aerosol sampler (Bright and Fletcher, 1983) IOM thoracic sampler (Vincent *et al.*, 1993)
7.4 L/min	Vertical elutriator (Görner *et al.*, 1994)
16.7 L/min	Dichotomous sampler (Loo *et al.*, 1976) Dzubay *et al.*, 1977; Loo *et al.*, 1979)
16.7 L/min	Monocut (John *et al.*, 1983)
< 20 L/min	Cascade impactors (Hering, 1995)
C. Medium	
113 L/min	Wedding 4 CFM ambient aerosol sampler (Wedding *et al.*, 1983)
113 L/min	McFarland (McFarland and Ortiz, 1982)
20–150 L/min	Cascade impactors (Hering, 1995)
D. High	
1133 L/min > 150 L/min	Size-selective inlet for Hi-Vol (McFarland *et al.*, 1984; Wedding and Weigand, 1985) Cascade impactors (Hering, 1995)

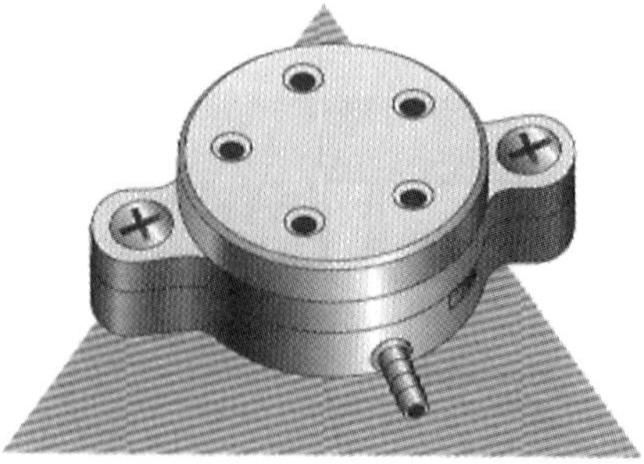

FIGURE 7.3. Personal environmental monitor (PEM) sampler from SKC Inc.

than about d_{ae} = 1 μm are not collected by the foam in the CIP10, resulting in a negative bias for aerosols containing fume-sized particles. Vincent and co-workers developed a combined inhalable/thoracic/respirable sampler using the inlet of the IOM sampler and separating the fractions using sections of porous plastic foam (Vincent *et al.*, 1993). The respirable dust is collected on a filter, while the larger fractions are measured by analyzing the dust collected on the foam sections. For the latter, it should be noted that each foam material being considered for use should be tested for moisture absorption prior to its use in this application. Kenny and Gussman (1997) developed a cyclone (see Figure 7.4) that collects the thoracic fraction at about 1.6 L/min or the respirable fraction at 4.2 L/min. The initial study estimated the flowrate to achieve the TPM curve by use of a model. Further experimentation is needed to obtain an accurate measurement of the optimal flowrate for TPM measurement. However, the performance curve for this cyclone appears to have a slope that matches the thoracic curve more accurately than most other cyclones, and Kenny and Gussman give an empirical equation for selecting cyclone dimensions for operating at other flowrates (for the same particle size fraction). Except for the IOM sampler, none of the personal thoracic samplers have been tested in wind tunnels to measure inlet efficiency at moderate to high windspeeds.

Person-wearable cascade impactors have been developed which can be used to sample the thoracic fraction. Ramachandran and Vincent (1997) discuss data inversion techniques for retrieving the inhalable, thoracic and respirable fractions from the raw data obtained using such instruments. The Andersen 290 sampler (Graseby-Andersen, Smyrna, GA), also known as the "Marple" (named after its originator) can be used with up to eight stages having cut points from d_{ae} = 0.5 to 20 μm (Marple and McCormack, 1983). The impaction stages each have six radial slots that are staggered relative to subsequent stages. This allows a jet plate to be the collection substrate for the previous stage. Two of the stages have cuts at d_{ae} = 9.8 and 3.5 μm respectively, allowing nearly direct measurement of the thoracic and respirable fractions. The overall design is compact and light and has been used extensively for measuring particle size distributions in the workplace. A similar device, the *personal inhalable dust spectrometer* (PIDS) (SKC Inc., Eighty-Four, PA), was developed in the UK for making personal exposure size distribution measurements (Gibson *et al.*, 1987). This device has so far been used to estimate inhalable, thoracic and respirable dust fractions in coal mines (Mark *et al.*, 1988), the primary lead production industry (Spear *et al.*, 1998), and elsewhere.

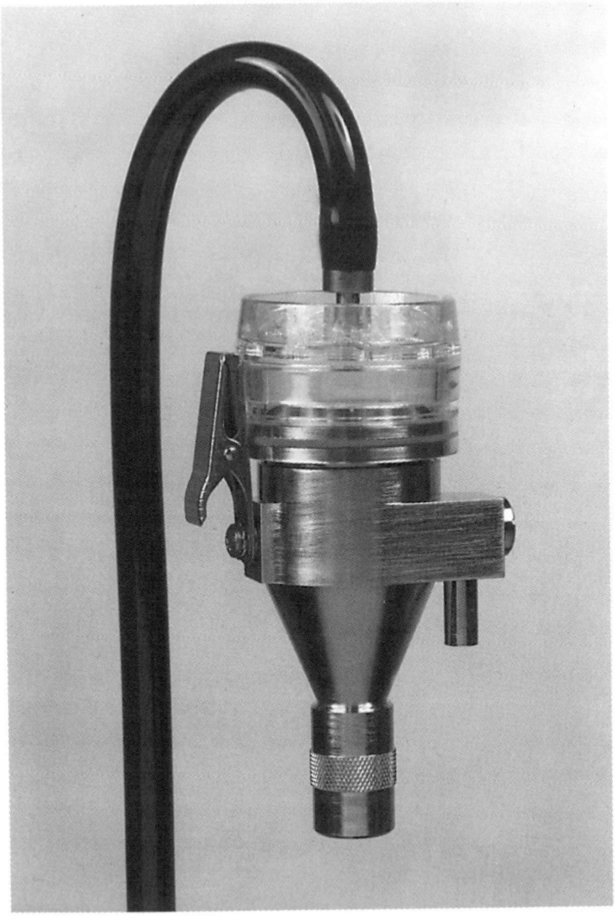

FIGURE 7.4. GK2.69 cyclone from BGI Inc. This device is designed to sample TPM at 1.6 L/min and RPM at 4.2 L/min.

A person-wearable cascade virtual impactor, called the "RespiCon," has been introduced (TSI Inc., St. Paul, MN) that collects IPM, TPM, and RPM simultaneously (see Figure 7.5). At present, however, no independent evaluations of this device have been published.

In the low volume category, the cotton dust vertical elutriator operating at 7.4 L/min was originally designed to have a 50% cut point at d_{ae} = 15 μm based on vertical flow in the body of the sampler. But due to jet formation at the inlet and internal turbulence, it has been found by separate researchers to have an actual measured 50% cut point at d_{ae} = 12 μm (Claassen, 1981) and 10.7 μm (Fabries *et al.*,

1989). This particular sampler has been used for over twenty years to estimate the exposures of cotton dust workers at risk from byssinosis.

Also in the low volume category, the dichotomous sampler (Loo *et al.*, 1979) is a virtual impactor having a flowrate of 16.7 L/min. The thoracic particulate mass fraction is selectively passed through the inlet; the virtual impaction stage further fractionates the aerosol into coarse and fine fractions with a cut point at d_{ae} = 2.5 μm.

A small portable sampler, operating at 6 L/min, has been developed by Bright and Fletcher (1983). The thoracic cut is provided by the inlet, which contains a single stage impactor with an oil-soaked porous plate to suppress particle bounce. A second particle size cut is made at d_{ae} = 3 μm by a Nuclepore filter (the sampler could be operated without it). The collection efficiency of the sampler's inlet, determined in a wind tunnel, is more sensitive to windspeed than the other samplers described above, but the performance may be adequate for relatively calm air or air which is slowly moving relative to the worker.

For thoracic aerosol mass sampling alone, the virtual impaction stage of the dichotomous sampler is unnecessary. The fractionating inlet can be coupled directly to a filter to form a sampler that has been called the "Monocut." Such a sampler using the earlier dichotomous sampler inlet with a d_{ae} = 15 μm cut point performed well (John *et al.*,

FIGURE 7.5. The "RespiCon" sampler for IPM, TPM and RPM from TSI, Inc., St. Paul.

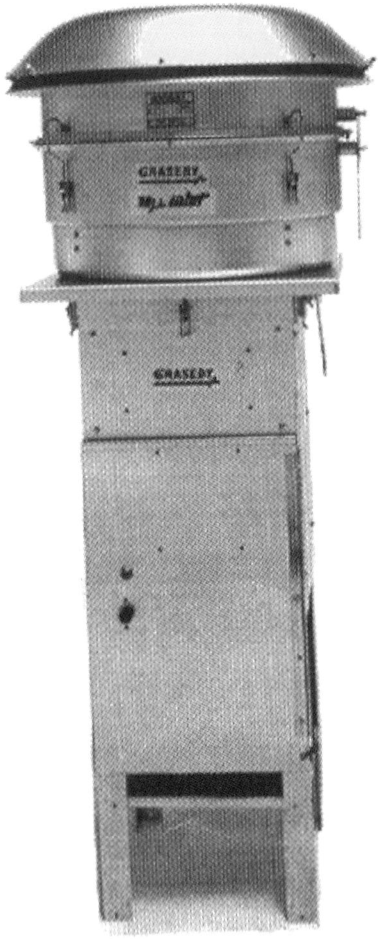

FIGURE 7.6. The high-volume PM_{10} sampler from Graseby-Andersen.

1983). The newer 10-μm cut point inlets should work equally well, affording an alternative to the dichotomous sampler.

Medium-volume samplers have been developed by McFarland and Ortiz (1982) and by Wedding *et al.* (1983). McFarland and Ortiz employed a sampler geometry that fractionates particles by a combination of impaction and sedimentation. The tortuous air path also suppresses particle bounce. A high volume sampler based on a similar geometry, called the "size-selective inlet" (SSI), converts a standard "hi-vol" sampler into a thoracic sampler (McFarland, *et al.*, 1979) (see Figure 7.6). The SSI can be used only with glass fiber or quartz filters.

FIGURE 7.7. The Wedding high-volume PM_{10} sampler.

The Andersen "Dichot" inlet (Liu and Pui, 1981; Shaw *et al.*, 1983) is based on the design of McFarland *et al.* (1978). Independence of wind direction is assured by cylindrical symmetry about the vertical axis. Wind tunnel tests verified that the dependence of the sampling efficiency of this device on wind direction was within the EPA-prescribed tolerances (EPA, 1987). Another inlet, developed by Wedding *et al.* (1982) (see Figure 7.7), is cylindrically symmetric about the vertical axis but employs a cyclonic action produced by turning vanes to achieve the thoracic fractionation. This inlet has also undergone wind tunnel testing with satisfactory results. A PM_{10} sampler using a dichotomous inlet with the capability of sequential sampling onto multiple filters is also commercially available. So too are real-time PM_{10} samplers using dichotomous inlets. One version, the "tapered element oscillating microbalance" (or "TEOM"), uses an inertial mass sensor (Patashnick and Rupprecht, 1991), another uses a beta gauge (Williams *et al.*, 1993).

In Figure 7.1, the measured sampling efficiencies of two of the samplers discussed above are compared to the TPM sampling criterion. For these, oleic acid particles were generated in wind tunnels operated at a windspeed of 0.56 m/s. The data points are seen to lie within the accepted range. These particular samplers were chosen for illustrative purposes only. A later chapter discusses the question of how to determine whether a sampler meets the performance criteria for TPM.

REFERENCES

Berry, R.D. and Froude, S. (1989), An Investigation of Wind Conditions in the Workplace to Assess Their Effect on the Quantity of Dust Inhaled, IR/L/DS/89/3, UK Health and Safety Executive, London.

Bright, D.S. and Fletcher, R.A. (1983), New portable ambient aerosol sampler, *Am. Ind. Hyg. Assoc. J.*, 44, pp. 528-536.

Buckley, T.J.,Waldman, J.M., Freeman, N.C.G. *et al.* (1991), Calibration, intersampler comparison and field application of a new PM-10 personal air-sampling impactor, *Aerosol Sci. Technol.*, 14, pp. 380-387.

Claassen, B.J. (1981), Experimental determination of the vertical elutriator particle transfer efficiency, *Am. Ind. Hyg. Assoc. J.*, 42, pp. 305-309.

Dzubay, T.C., Stevens, R.L. and Peterson, C.M. (1977), In: *X-Ray Fluorescence Analysis of Environmental Samples* (T.C. Dzubay, Ed.), Ann Arbor Science Publishers, Ann Arbor, MI.

Fabriés, J.F., Görner, P. and Wrobel, R. (1989), A new air sampling instrument for the assessment of the thoracic fraction of an aerosol, *J. Aerosol Sci.*, 20, pp. 1589-1592.

Gibson, H., Vincent, J.H. and Mark, D. (1987), A personal inspirable aerosol spectrometer for applications in occupational hygiene research, *Ann. Occup. Hyg.*, 31, pp. 463-479.

Görner, P., Fabriés, J.-F. and Wrobel, R. (1994), Thoracic fraction measurement of cotton dust, *J. Aerosol Sci.*, 25 (Suppl. 1), pp. S487-S488.

Hering, S.V. (1995), In: *Air Sampling Instruments for Evaluation of Atmospheric Contaminants* (B.S. Cohen and S.V. Hering, Eds.), American Conference of Governmental Industrial Hygienists (ACGIH), Cincinnati, OH, Chapter 14.

John, W. (1993), Instrument performance and standards for sampling of aerosols. *Appl. Occup. Environ. Hyg.*, 8, pp. 251-259.

John, W. and Reischl, G. (1978), Measurement of the filtration efficiencies of selected filter types, *Atmos. Environ.*, 12, pp. 2015-2019.

John, W., Wall, S.M. and Wesolowski, J.J. (1983), *Validation of Samplers for Inhaled Particulate Matter, EPA-600/S4-83-010, Environmental Protection Agency.*

John, W., Wirklmayr, W., Wang, H.C.: (1991), Particle deagglomeration and reentrainment in a PM_{10} sampler, *Aerosol Sci. Technol.*, 14, pp. 165-176.

Kenny, L.C. and Gussmann, R.A. (1997), Characterization and modeling of a family of cyclone aerosol preseparators, *J. Aerosol Sci.*, 28, pp. 677-688.

Lippmann, M. (1995), In: *Air Sampling Instruments for Evaluation of Atmospheric Contaminants* (B.S. Cohen and S.V.Hering, Eds.), American Conference of Governmental Industrial Hygienists (ACGIH), Cincinnati, OH, Chapter 13.

Liu, B.Y.H. and Pui, D.Y.H. (1981), Aerosol sampling inlets and inhalable particles, *Atmos. Environ.*, 15, pp. 589-600.

Loo, B.W., Jaklevic, J.M. and Goulding, F.S. (1976), In: *Fine Particles: Aerosol Generation, Measurement, Sampling and Analysis* (B.Y.H. Liu, Ed.), New York, Academic Press, pp. 311-350.

Loo, B.W., Adachi, R.S. and Cork, C.P. (1979), A second-generation dichotomous sampler for large-scale monitoring of airborne particulate matter, LBL-8725. (Lawrence Berkeley Laboratory, Berkeley, CA).

Mark, D., Cowie, H., Vincent, J.H. *et al.* (1988), The variability of exposure of coalminers to inspirable dust, Report TM/88/0, Institute of Occupational Medicine, Edinburgh, U.K.

Marple, V.A.(1989), PEM development, fabrication, evaluation and calibration, Subcontract number 3-321U-3993 Work Assignment No. 34., MSP Corporation.

Marple, V.A. and McCormack, J.E. (1983), Personal sampling impactor with respirable aerosol penetration characteristics, *Am. Ind. Hyg. Ass. J.*, 44, pp. 916-922.

McFarland, A.R. and Ortiz, C.A. (1982), A 10-μm cut point ambient aerosol sampling inlet, *Atmos. Environ.*, 16, pp. 2959-2965.

McFarland, A.R., Oritz, C.A. and Bertch, R.W.J.: (1984), 10 μm cut point size selective inlet for hi-volume samplers, *J. Air Pollut. Control Assoc.*, 34, p. 544.

McFarland, A.R., Ortiz, C.A. and Bertsch, R.W.J. (1978), Particle collection characteristics of a single-stage dichotomous sampler, *Environ. Sci. Technol.*, 12, p. 679.

McFarland, A.R., Ortiz, C.A. and Bertsch, R.W.J. (1979), A high-capacity pre-separator for collecting large particles, *Atmos. Environ.*, 13, p. 761.

Patashnick, H. and Rupprecht, E.G. (1991), Continuous PM-10 measurements using the tapered element oscillating microbalance, *J. Air Waste Manage. Ass.*, 41, pp. 1079-1083.

Ramachandran, G. and Vincent, J.H. (1997), Evalution of two inversion techniques for retrieving health-related aerosol fractions from personal cascade impactor measurements, *Am. Ind. Hyg. Ass. J.*, 58, pp. 15-22.

Rao, A.K. and Whitby, K.T. (1977), Non-ideal collection characteristics of single-stage and cascade impactors, *Am. Ind. Hyg. Ass. J.*, 38, pp. 174-179.

Rubow, K.L., Marple, V.A., Olin, J. and McCawley, M.A. (1987), A personal cascade impactor: design, evaluation and calibration, *Am. Ind. Hyg. Ass. J.*, 48, pp. 532-538.

Shaw, R.W., Stevens, R.K., Lewis, C.W. *et al.* (1983), Comparison of aerosol sampler inlets, *Aerosol Sci. Technol.*, 2, pp. 53-68.

Spear, T.M., Werner, M.A., Bootland, J. *et al.* (1998). Assessment of particle size distributions of health-relevant aerosol exposures of primary lead smelter workers, *Ann. Occup. Hyg.*, 42, pp. 73-80.

Tufto, P. and Willeke, K. (1982), Dependence of particulate sampling efficiency on inlet orientation and flow velocities, *Am. Ind. Hyg. Ass. J.*, 43, pp. 436-443.

U.S. Environmental Protection Agency (EPA) (1987), Revisions to the national ambient air quality standard for particulate matter, *Federal Register*, 52(July 1), pp. 24634-24750.

Vincent, J.H., Aitken, R.J. and Mark, D. (1993), Porous plastic foam filtration media: penetration characteristics and applications in particle size-selective sampling. *J. Aerosol Sci.*, 24, pp. 929-944.

Wedding, J.B. (1982), Ambient aerosol sampling: history, present thinking, and a proposed inlet for inhalable particles, *Environ. Sci. Technol.*, 16, pp. 154-161.

Wedding, J.B. and Weigand, M.A.: (1985), The Wedding Ambient Aerosol Sampling Inlet (D_{50} = 10 μm) for the High-Volume Samplers, *Atmos. Environ.*, 19, p. 535.

Wedding, J.B., McFarland. A.R. and Cermak, J.E. (1977), Large particle collection characteristics of ambient aerosol samplers, *Environ. Sci. Technol.*, 11, pp. 387-390.

Wedding, J.B., Weigand, M., John, W. *et al.* (1980), Sampling effectiveness of the inlet to the dichotomous sampler, *Environ. Sci. Technol.*, 14, pp. 1367-1370.

Wedding, J.B., Weigand, M. and Carney, T.C. (1982), A 10-μm cut point inlet for the dichotomous sampler, *Environ. Sci. Technol.*, 18, pp. 602-606.

Wedding, J.B., Weigand, M.A. and Ligotke, M.W. (1983), The Wedding ambient aerosol sampling inlet for an intermediate flowrate (4 cfm) sampler, *Environ. Sci. Technol.*, 17, pp. 379-383.

Wesolowski, J.J., John, W. and Devor, W. (1977), In: *X-Ray Fluorescence Analysis of Environmental Samples* (Dzubay, T.G., Ed.), Ann Arbor Science Publishers, Inc., Ann Arbor, MI, pp. 121-131.

Williams, K., Fairchild, C. and Jaklevic, J. (1993), In: *Aerosol Measurement: Principles, Techniques, and Applications* (K. Willeke and P.A. Baron, Eds.), New York, Van Nostrand Reinhold, pp. 296-312.

Chapter 8

SAMPLING FOR RESPIRABLE AND FINE AEROSOL

Walter John

Particle Science, Walnut Creek, CA

8.1 General Requirements for the Ideal Sampler

Most of the general requirements for a respirable aerosol sampler are similar to those for the thoracic aerosol sampler except for characteristics affected by the difference in the particle size range. The particle size should be selected aerodynamically, following the respirable particulate matter (RPM) criterion presented in Chapter 4. Because the aerodynamic diameters of the particles to be sampled are smaller than those for the other health-related fractions, the inlet requirements are generally less stringent for respirable than for thoracic sampling. For example, the RPM criteria specify a sampling efficiency of only 1.3% at $d_{ae} = 10$ µm where the thoracic particulate matter (TPM) criteria call for a 50% efficiency. Filter media selection criteria are essentially the same as those discussed for thoracic sampling. The concerns about leak-tight filter holders also apply to respirable sampling.

Respirable aerosol samplers should be equipped with flow controllers to maintain the flowrate constant to within 5% as the pressure drop across the filter increases and if the environmental temperature changes. Personal samplers usually employ a diaphragm pump, necessitating the use of a pulsation dampener, since pulsation has been shown to seriously affect the collection efficiency of size-selective samplers such as the 10-mm nylon cyclone which has long been familiar to industrial hygienists (Blachman and Lippmann, 1974;

Caplan *et al.*, 1977; Bartley *et al.*, 1984). There is evidence that electrostatic charge too affects the performance of nylon cyclones (Almich and Carson, 1974; Briant and Moss, 1984). Therefore, in general, the use of samplers made with conductive materials is to be recommended.

Although respirable aerosol samplers are somewhat easier to design than thoracic samplers, they still require thorough testing and validation, including any modifications. Again, the user must base the choice of sampler on performance criteria with the particular application in mind.

8.2 SAMPLING DEVICES FOR RESPIRABLE AEROSOL

The salient features of the principal types of respirable samplers will be reviewed here. Specific samplers are listed in Table 8.1. A more comprehensive review has been presented elsewhere by Lippmann (1995).

CYCLONES

Cyclones are probably the most commonly used samplers for respirable aerosol, at least as personal samplers. They are available in a wide range of flowrates, including a small size operating at relatively low flowrates so that a miniature pump suitable for personal sampling can be used. In general for cyclones, the sampling efficiency can be fairly closely matched to that of the RPM curve. It has been pointed out, however, that the collection efficiency is sensitive to the outlet configuration (Lippmann, 1995). Cyclones have important practical advantages, including minimal particle bounce and reentrainment problems, large capacity for loading, and insensitivity to orientation. A disadvantage of the cyclone is a lack of a fundamental theory which can predict performance. However, empirical theories are available to assist the designer (Chan and Lippmann, 1977; John and Reischl, 1980; Saltzman and Hochstrasser, 1983; Lidén and Gudmundsson, 1997).

Considerable data are available on the performance of the 10-mm nylon cyclone (Ettinger *et al.*, 1970; Baron, 1983). In Figure 8.1, the most recent data from Bartley *et al.* (1994) are shown for four different flowrates together with the curve following the RPM sampling criterion. For a flowrate of 1.7 L/min, selected as optimum, a bias map is shown in Figure 8.2. Any point on this diagram represents an aerosol with a given particle size distribution and gives the percentage differ-

TABLE 8.1. Respirable particulate matter (RPM) samplers

A. Cyclones	10 mm Nylon (Caplan *et al.*, 1977; Bartley *et al.*, 1994), ASL (Lippmann and Chan 1970), Aerotec (Lippmann and Chan, 1970 and 1979), Unico (Lippmann and Chan, 1979), AIHL (John and Reischl, 1980), others.
B. Elutriators	Horizontal: MRE (Dunmore *et al.*, 1964), Hexhlet (Wright, 1954), Vertical: LSG (Vekeny, 1971).
C. Impactors	Andersen (Andersen, 1958), Lundgren (Lundgren, 1967), Marple (Marple and Rubow, 1983; Marple and McCormack, 1983), MOUDI (Hering, 1995), Berner (Berner and Lurzer, 1980), others.
D. Virtual impactors	Dichotomous sampler (Loo *et al.*, 1976 and 1979; Dzubay *et al.*, 1977), Centripeter (Hounam and Sherwood, 1965).
E. Filters	Filter pack (Reiter and Potzl, 1967), porous plastic foam filter (Gibson and Vincent, 1981; Vincent *et al.*, 1993), Nuclepore filter (Cahill *et al.*, 1977; Parker *et al.*, 1977; Heidem, 1981; John *et al.*, 1983)

Note: Commercially-available samplers are described in Hering (1995) and elsewhere.

ence between the mass which would be sampled by the 10-mm nylon cyclone and that which would be sampled by a hypothetical ideal RPM sampler. In this figure, the particle size distribution is assumed to be a lognormal function having a mass median aerodynamic diameter

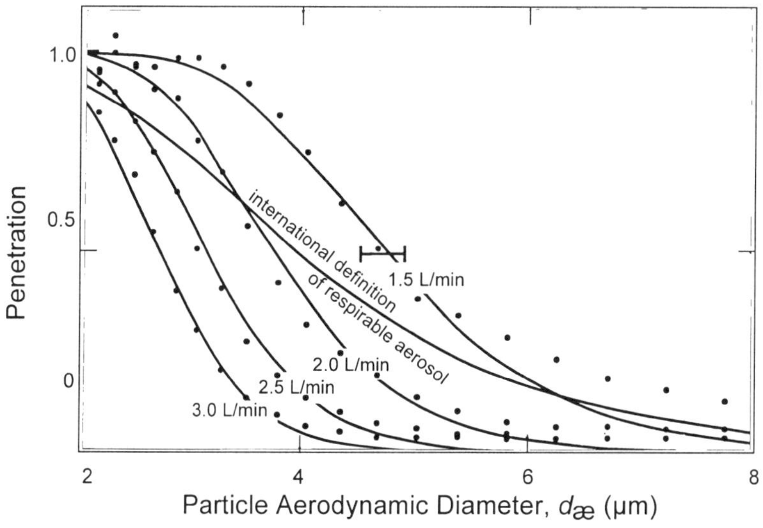

FIGURE 8.1. Sampling efficiency of the 10-mm nylon cyclone for four flowrates and the curve following the respirable criteria (Bartley *et al.*, 1994).

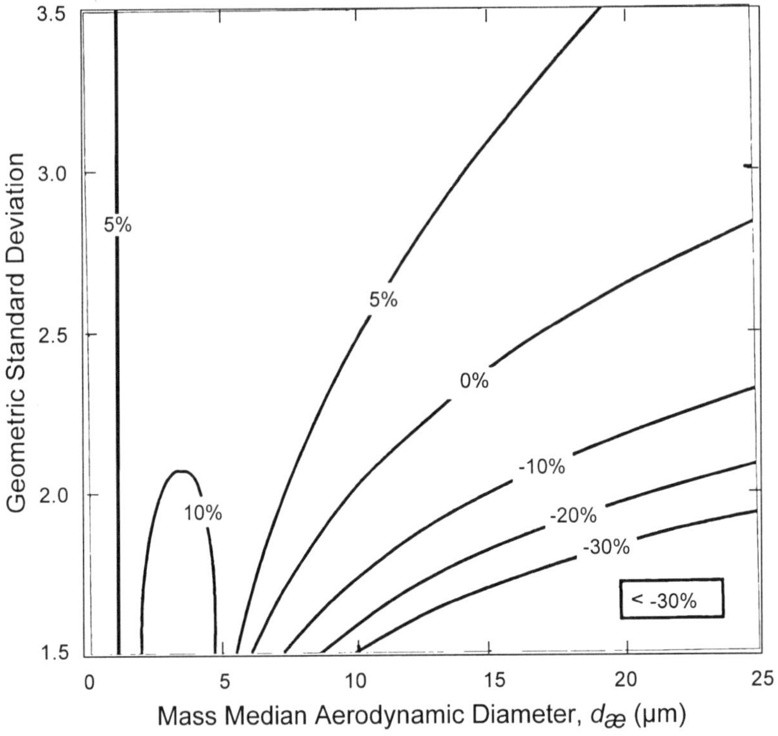

FIGURE 8.2. Bias map of the 10-mm nylon cyclone at 1.7 L/min (Bartley *et al.*, 1994).

corresponding to the x-coordinate and a geometric standard deviation corresponding to the y-coordinate.

Lidén and Gudmundsson (1996) have optimized the configuration and dimensions of a cyclone to match the sampling efficiency to that of the current RPM curve. The performance of a new respirable multi-inlet cyclone has been reported by Gautam and Sreenath (1998). Kenny and Gussman (1997) have characterized a family of cyclones and identified those sampling according to the RPM criterion. John (1997) has developed a personal respirable aerosol sampler using a spiral channel as a particle size-selector. This sampler has sampling characteristics similar to those of a cyclone including the penetration curve and the absence of particle bounce.

Although, as stated earlier, inlet losses are generally less problematical than for the coarser fractions (i.e., IPM and TPM), some problems have been noted at the high windspeeds pertaining to some working environments (e.g., mining), where sampling efficiency for the respirable fraction may be quite strongly influenced by the effects of the external wind at the sampler entry (Cecala *et al.*, 1983).

HORIZONTAL ELUTRIATORS

Horizontal elutriators have been widely used, particularly in Britain and elsewhere in Europe, especially in the mining industry (Dunmore *et al.*, 1964). The main advantage of this type of pre-selector is its predictable and deterministic performance based on gravitational settling of the particles during passage between horizontal collecting plates (Walton, 1954). In fact, it was knowledge of this performance that provided the basis for the earliest convention for the respirable aerosol fraction (i.e., the BMRC curve, see Chapter 4). Disadvantages include the restriction to a fixed orientation, possible reentrainment of particle deposits and the difficulty of miniaturization. Similar to the cyclone, some dependence of performance on windspeed has been noted at high windspeeds (e.g., like those encountered in mining) (Ogden *et al.*, 1977).

IMPACTORS

The collection efficiency of an impactor can be accurately predicted from numerical calculations based on theory (Marple and Liu, 1974; Marple and Willeke, 1976). On the other hand, important details such as wall losses cannot be reliably predicted. Therefore, as for all other samplers, thorough testing is required. Impactors can be designed over a wide range of flowrates and can be operated at any orientation. A further advantage of the impactor is the possibility of cascading stages to provide particle size-segregated samples which can be analyzed to produce the particle size distributions of specific chemical species (as in the "cascade impactor").

Impactors suffer from two inherent drawbacks, particle bounce (Rao and Whitby, 1977; Wesolowski *et al.*, 1977) from the impaction surface and reentrainment because of the continuous blowing of the jet on the particle deposit. The nonuniformity of the deposit is also disadvantageous for some analyses. Various strategies have been employed to suppress particle bounce, including the use of appropriate filter substrates. Here, however, because the aerosol jet partially penetrates the filter surface, some small particles are lost, degrading the performance (Rao and Whitby, 1977). The use of grease coatings has been found to be effective, provided that the grease itself is not allowed to become excessively coated with particles. This in turn imposes stringent limitations on the permissible loading.

For an impaction stage which acts as a pre-selector, allowing the respirable fraction to penetrate to an after-filter, Reischl and John (1978) showed that an oil-soaked, porous plate can tolerate very high loadings. Marple and Rubow (1983) have employed this technique in their respirable aerosol impactor which features a combination of variously sized jet nozzles to produce a stair-step approximation of the respirable curve. A so-called "*universal impactor*" was developed by John (1995), featuring a tapered slit which yields a penetration curve closely following the RPM criterion.

A fundamental approach to overcome the particle bounce problem is the use of the "*virtual impactor*" concept. Although early versions of the virtual impactor (Conner, 1956; Hounam and Sherwood, 1965) suffered from poor performance, including high wall losses, the sampler development program sponsored by the U.S. Environmental Protection Agency (EPA) (Miller *et al.*, 1979) has since produced good-performance, well-tested virtual impactors with cut points at 3.5 µm or 2.5 µm (Loo *et al.*, 1976 and 1979; Dzubay *et al.*, 1977). It should be emphasized, however, that the cut-off curves for all impactors, including virtual impactors, are generally relatively sharp and hence do not conform closely to the shape of the RPM curve. That notwithstanding, a new instrument has recently appeared, the "RespiCon" (Dunkhorst *et al.*, 1995), which utilizes virtual impaction to achieve the appropriate shape of penetration curve and which can provide not only the respirable fraction corresponding to the RPM curve but also the inhalable (IPM) and thoracic (TPM) fractions.

FILTERS

Inefficient filters have been used as pre-collectors, allowing the respirable fraction of particles to penetrate to an after-filter. Porous plastic foam media have been used for this purpose and have been shown to produce a penetration curve whose shape is very close to the RPM curve (Gibson and Vincent, 1981; Vincent *et al.*, 1993). Large-pore Nuclepore filters have also been used as such pre-collectors (Cahill *et al.*, 1977; Parker *et al.*, 1977; Heidem, 1981). Here correspondence to the RPM curve is only approximate because the particle size-selection is primarily by interception and hence is determined by the geometrical size of the particles (John *et al.*, 1983). Since the particle aerodynamic diameter increases with the square root of the particle density, a serious error will occur in the sampling of high density materials such as heavy metals. In addition, there is also a serious particle bounce problem, although this can be mitigated — but not eliminated — by the application of a grease coating. Another problem of the Nuclepore filters is the very low capacity for particle loading. The 'sequential' Nuclepore filter sampler is particularly simple and low in cost and therefore may be useful in certain situations, but it is not recommended for general use.

8.3 Sampling for $PM_{2.5}$

In 1997, the EPA revised the National Ambient Air Quality Standards for particulate matter in order to increase protection against particulate-related health effects (EPA, 1997a). The new standards are based on the sampling of $PM_{2.5}$, where $PM_{2.5}$ refers to particulate matter with a particle size generally smaller than a nominal 2.5 μm in aerodynamic diameter. (The precise EPA definition of $PM_{2.5}$ is given below.) The rationale for choosing a 2.5-μm size cut derives from the characteristics of atmospheric particle size distributions, which can be fitted by multiple lognormal modes (EPA, 1997b). There is usually a minimum at about 2.5 μm between the fine particle accumulation mode and the coarse particle mode (John, 1993). The fine particle fraction is found to consist mainly of combustion-derived particles (mostly generated by human activities), whereas the coarse fraction consists primarily of wind-blown mineral particles. A 2.5-μm cut serves to separate the fine and coarse fractions which have different physical and chemical characteristics, as well as different origins. Further, it is believed that health effects are more likely to be associated with the fine particles.

$PM_{2.5}$ is defined by the EPA (1997b) as the mass of particles collected on a Teflon filter by a reference sampler. The reference sampler is completely specified from the inlet down to and including the filter. The specifications include dimensions of the mechanical components, tolerances, materials and surface finishes. The remainder of the sampler, pump, flow control, data storage, temperature and pressure sensors, etc., are specified by performance standards. The inlet for the reference sampler is shown in Figure 8.3. This inlet is, in fact, the inlet of the dichotomous sampler, with a slight modification to the vanes around the top to reduce the intake of precipitation. The inlet assembly includes a size-selective stage which provides a pre-cut at 10 μm. The aerosol then enters an impaction stage, shown in Figure 8.4, which provides the 2.5-μm cut. The impaction surface, located in a well, is covered with a glass fiber filter which, in turn, is covered by a thin layer of low-volatility oil to prevent particle bounce and reentrainment. This impactor configuration produces a particle size cut-off which is not as sharp as that of a classical impactor. The cut-off curve

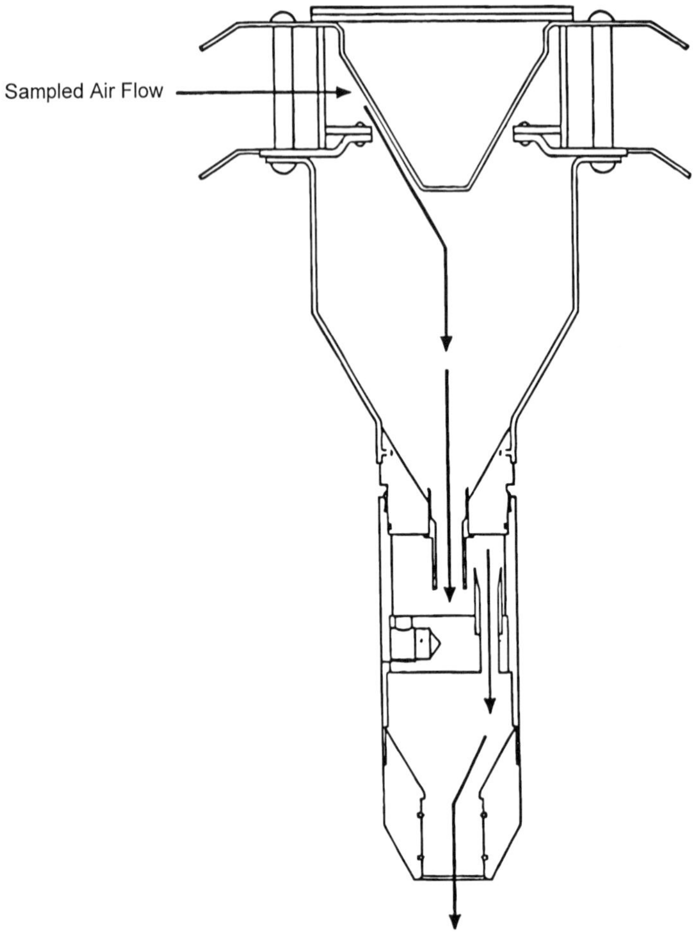

FIGURE 8.3 Inlet for the $PM_{2.5}$ Federal Reference Method sampler (EPA, 1997b).

is shown in Figure 8.5. The $PM_{2.5}$ reference sampler is currently available from several manufacturers.

The EPA has specified several classes of equivalent methods for $PM_{2.5}$ (EPA, 1997b). Candidate Class I samplers are those deviating in only a minor way from the reference sampler. Only the modifications are required to be tested. Candidate Class II samplers deviate substantially from the reference method and must undergo extensive, stringent testing. All other samplers fall into the Class III category. These include continuous monitors such as beta gages and the TEOM, and non-filter samplers such as nephelometers and other optical devices. Class III testing is handled on a case-by-case basis.

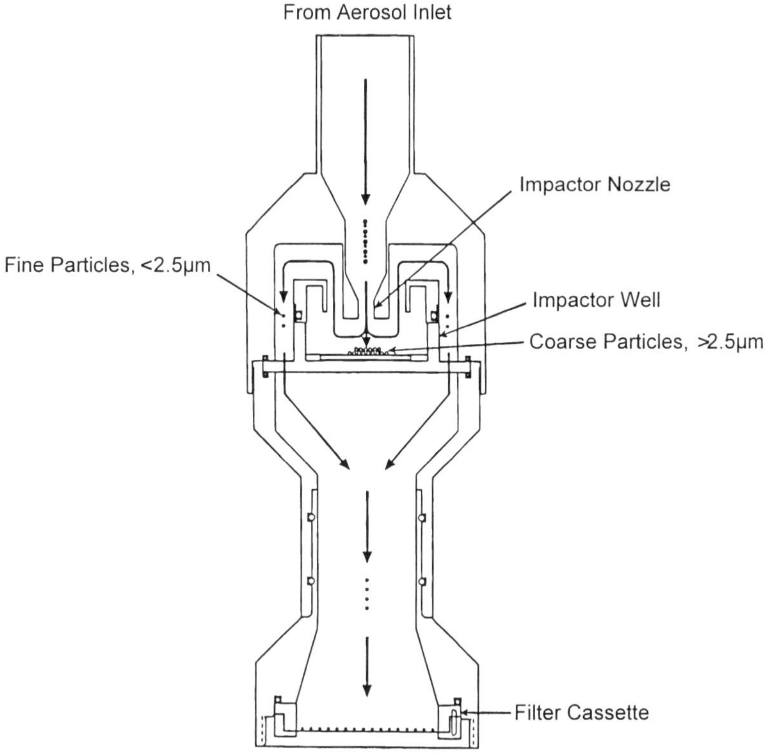

FIGURE 8.4 Fractionation stage and filter assembly for the $PM_{2.5}$ Federal Reference Method sampler (EPA, 1997b).

The EPA-required testing of the sampler fractionator for Class II equivalency involves measuring the sampling efficiency versus particle size and from the resulting efficiency curve, calculating the mass which would be sampled for three different specified particle size distributions. The calculated mass is to be within 5% of that calculated for the EPA's ideal $PM_{2.5}$ sampler. Additionally, the cut point is to be 2.5 ± 0.2 µm.

The new $PM_{2.5}$ standards involve only collected particulate mass. In recognition of the need to obtain data on the chemical composition of $PM_{2.5}$, the EPA is planning a special sampling network (EPA, 1997b). In general, the reference sampler is not appropriate for this purpose. A newly-available sampler specifically designed for this application, is the Spiral Aerosol Speciation Sampler (SASS), which has a $PM_{2.5}$ spiral inlet (John, 1997), a denuder, and two sequential filters. Other samplers for speciation of $PM_{2.5}$ are under development by several manufacturers.

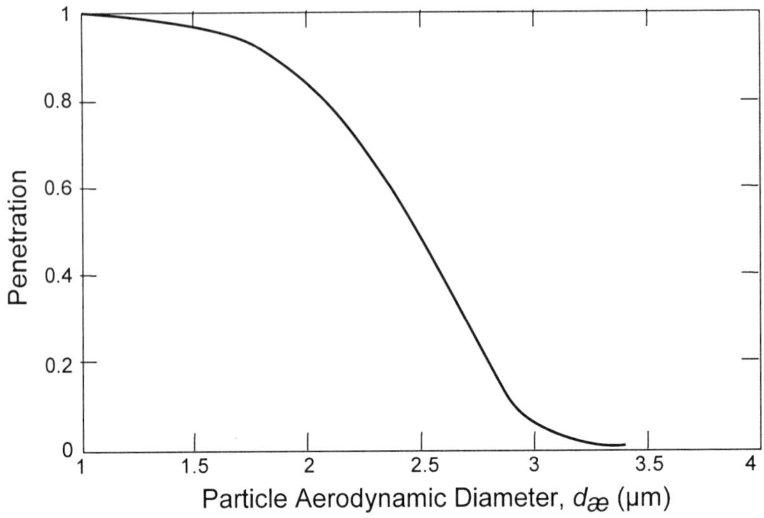

FIGURE 8.5 Sampling efficiency versus particle aerodynamic diameter for the $PM_{2.5}$ Federal Reference Method sampler (EPA, 1997a).

For personal sampling of $PM_{2.5}$, the Personal Environmental Monitor (PEM) (Hering, 1995) is a lapel-worn sampler having a single-stage impactor followed by a final filter (see Figure 7.3). The impaction stage is oiled to prevent bounce. The 2.5 µm cut is sharp. The Spiral $PM_{2.5}$ Particle Sampler is also a lapel-worn sampler. A spiral channel (John, 1997) provides a cutoff curve closely fitting the EPA ideal $PM_{2.5}$ curve.

REFERENCES

Almich, B.P. and Carson, G.A. (1974), Some effects of charging on 10-mm Nylon cyclone performance, *Am. Ind. Hyg. Ass. J.*, 34, pp. 603-612.

Andersen, A.A. (1958), A new sampler for collecting, sizing, and enumeration of viable airborne bacteria, *J. Bact.*, 76, p. 471.

Baron, P.A. (1983), Sampler evaluation with an aerodynamic particle sizer, In:.*Aerosols in the Mining and Industrial Work Environments* (V.A. Marple and B.Y.H. Liu, Eds.), Ann Arbor Science, Ann Arbor, MI, Chapter 61.

Bartley, D.L., Breuer, G.M., Baron, P.A. *et al.* (1984), Pump fluctuations and their effect on cyclone performance, *Am. Ind. Hyg. Ass. J.*, 45, pp. 10-18.

Bartley, D.L., Chen, C-C, Song, R. *et al.* (1994), Respirable aerosol performance testing, *Am. Ind. Hyg. Ass. J.*, 55, pp. 1036-1046.

Berner, A. and Lurzer, C. (1980), Mass size distributions of traffic aerosols at Vienna, *J. Phys. Chem.*, 84, pp. 2079-2083.

Blachman, M.W. and Lippmann, M. (1974), Performance characteristics of the multicyclone aerosol sampler, *Am. Ind. Hyg. Ass. J.*, 35, pp. 311-326.

Briant, J.K. and Moss, O.R. (1984), The influence of electrostatic charges on the performance of the 10-mm nylon cyclone, *Am. Ind. Hyg. Ass. J.*, 45, pp. 440-445.

Cahill, T.A., Ashbaugh, L.L., Barone, J.B. *et al.* (1977), Analysis of respirable fractions in atmospheric particulates via sequential filtration, *J. Air Poll. Contr. Ass.*, 77, pp. 675-679.

Caplan, K.J., Doemeny, L.J. and Sorenson, S.D. (1977), Performance characteristics of the 10-mm cyclone respirable mass sampler: Part 1 Monodisperse studies, *Am. Ind. Hyg. Ass. J.*, 38, pp. 83-95.

Cecala, A.B., Volkwein, J.C., Timko, R.J. *et al.* (1983), Velocity and orientation effects on the 10-mm Dorr-Oliver cyclone,U.S. Bureau of Mines Report of Investigations, RI 8764.

Chan, T.L. and Lippmann, M. (1977), Particle collection efficiencies of air sampling cyclones: an empirical theory, *Environ. Sci. Technol.*, 11, pp. 377-382.

Conner, W.D. (1956), An inertial-type particle separator for collecting large samples, *J. Air Poll. Control Ass.*, 16, p. 35.

Dunkhorst, W., Lodding, H. and Koch, W. (1995), A new personal monitor for real-time measurement of the inspirable, the thoracic, and the respirable fraction of airborne dust, *J. Aerosol Sci.*, 26, pp. S107-S108.

Dunmore, J.H., Hamilton, R.J. and Smith, D.S.G. (1964), An instrument for the sampling of respirable dust for subsequent gravimetric assessment, *J. Sci. Instr.*, 41, pp. 669-672.

Dzubay, T.G., Stevens, R.K. and Peterson, C.M. (1977), Application of the dichotomous sampler to the characterization of ambient aerosols, In: *X-Ray Fluorescence Analysis of Environmental Samplers* (T.G. Dzubay, Ed.), Ann Arbor Science Publishers, Ann Arbor, MI, Chapter 7. 77).

Ettinger, H.J., Partridge, J.E. and Royer, G.W. (1970), Calibration of two-stage air samplers, *Am. Ind. Hyg. Ass. J.*, 31, pp. 537-545.

Gautam, M. and Sreenath, A. (1998), Performance of a respirable multi-inlet cyclone, *J. Aerosol Sci.*, 28, pp. 1265-1281.

Gibson, H. and Vincent, J.H. (1981), The penetration of dust through porous foam filter media, *Ann. Occup. Hyg.*, 24, pp. 205-215.

Heidem, N.Z. (1981), Review: aerosol fractionation by sequential filtration with nuclepore filters, *Atmos. Environ.*, 15, pp. 891-904.

Hering, S.V. (1995), In: *Air Sampling Instruments for Evaluation of Atmospheric Contaminants*, 8th Edition (B.S. Cohen and S.V. Hering, Eds.), American Conference of Governmental Industrial Hygienists (ACGIH), Cincinnati, OH, Chapter 14.

Hounam, R.F. and Sherwood, R.J. (1965), The cascade centripeter: a device for determining the concentration and size distribution of aerosols, *Am. Ind Hyg. Ass. J.*, 26, p. 122.

John, W. (1993), The Characteristics of Environmental and Laboratory-Generated Aerosols, In: *Aerosol Measurement* (K. Willeke and P.A. Baron, Eds.), Van Nostrand and Reinhold, N.Y., Chapter 5.

John, W. (1995), *Universal impactor for particle collection within sampling criteria,* U.S. Patent No. 5,437,198, Aug.1, 1995, U.S. Patent Office.

John, W. (1997), New personal samplers for respirable particles and PM-2.5, Annual Conference of the American Association for Aerosol Research, Denver, CO, October 13 – 17. SKC, Inc., Eighty Four, PA.

John, W. and Reischl, G. (1980), A cyclone for size-selective sampling of ambient air. *J. Air Poll. Cont. Ass.*, 30, pp. 872-876.

John, W., Hering, S.V., Reischl, G. *et al.* (1983), Characteristics of nuclepore filters with large pore size: II Filtration properties, *Atmos. Environ.*, 17, pp. 373-382.

Kenny, L.C. and Gussman, R.A. (1997), Characterization and modeling of a family of cyclone aerosol pre-separators, *J. Aerosol Sci.*, 28, pp. 677-688.

Lidén, G. and Gudmundsson, A. (1996), Optimization of a cyclone to the 1993 international sampling convention for respirable dust, *Appl. Occup. Environ. Hyg.*, 11, pp. 1398-1408.

Lidén, G. and Gudmundsson, A. (1997), Semi-empirical modeling to generalise the dependence of cyclone collection efficiency on operating conditions and cyclone design, *J. Aerosol Sci.*, 28, pp. 853-874.

Lippmann, M. (1995), Size-selective health hazard sampling, In: *Air Sampling Instruments for Evaluation of Atmospheric Contaminants, 8th Edition* (B.S. Cohen and S.V. Hering, Eds.), American Conference of Governmental Industrial Hygienists (ACGIH), Cincinnati, OH, Chapter 5.

Lippmann, M. and Chan, T.L. (1970), Calibration of dual-inlet cyclones for respirable mass sampling, *Am. Ind Hyg. Ass. J.*, 31, pp. 189-200.

Lippmann, M. and Chan, T.L. (1979), Cyclone sampler performance, *Staub Reinhalt. Luft*, 39, pp. 7-11.

Loo, B.W., Jaklevic, J.M. and Goulding, F.S. (1976), Dichotomous virtual impactors for large-scale monitoring of airborne particulate matter, In: *Fine Particles, Aerosol Generation, Measurement, Sampling and Analysis* (B.Y.H. Liu, Ed.), Academic Press, Inc., New York, pp. 311-350.

Loo, B.W., Adachi, R.S., Cork, C.P. *et al.* (1979), A second-generation dichotomous sampler for large-scale monitoring of airborne particulate matter, LBL-8725. Lawrence Berkeley Laboratory, Berkeley, CA.

Lundgren, D.A. (1967), An aerosol sampler for determination of particle concentrations as a function of size and time, *J. Air Poll. Contr. Ass.*, 17, p. 4.

Marple, V.A. and Liu, B.Y.H. (1974), Characteristics of laminar jet impactors, *Environ. Sci. Technol.*, 8, pp. 648-654.

Marple, V.A. and McCormack, J.E. (1983), Personal sampling impactor with respirable aerosol penetration characteristics, *Am. Ind. Hyg. Ass. J.*, 44, 916-922.

Marple, V.A. and Rubow, K.L. (1983), Impactors for respirable dust sampling, In: *Aerosols in the Mining and Industrial Work Environments* (V.A. Marple and B.Y.H. Liu, Eds.), Ann Arbor Science, Ann Arbor, MI, Chapter 60.

Marple, V.A. and Willeke, K. (1976), Impactor design, *Atmos. Environ.*, 10, pp. 891-896.

Miller, F.J., Gardner, D.E., Graham, J.A. *et al.* (1979), Size considerations for establishing a standard for inhalable particles, *J. Air Poll. Contr. Ass.*, 29, pp. 610-615.

Ogden, T.L., Birkett, J.L. and Gibson, H. (1977), Improvements to dust measuring techniques, IOM Report No. TM/77/11, Institute of Occupational Medicine, Edinburgh, Scotland, U.K.

Parker, R.D., Buzzard, G.H., Dzubay, T.G. *et al.* (1977), A two-stage respirable aerosol sampler using nuclepore filters in series, *Atmos. Environ.*, 11, pp. 617-621.

Rao, A.K. and Whitby, K.T. (1977), Non-ideal collection characteristics of single-stage and cascade impactors, *Am. Ind. Hyg. Ass. J.*, 38, pp. 174-179.

Reischl, G. and John, W. (1978), The collection efficiency of impaction surfaces: a new impaction surface, *Staub-Reinhalt. Luft*, 38, p. 55.

Reiter, R. and Potzl, K. (1967), The design and operation of a respiratory tract model, *Staub Reinhalt. Luft*, 27, p. 19.

Saltzman, B. and Hochstrasser, J.M. (1983), Design and performance of miniature cyclones for respirable aerosol sampling, *Environ. Sci. Technol.*, 17, pp. 418-424.

U.S. Environmental Protection Agency (1997a), Draft, Quality Assurance Guidance Document 2.12 (Nov. 1997), Human Exposure and Atmospheric Sciences Division, National Exposure Research Laboratory, Research Triangle Park, NC 27711.

U.S. Environmental Protection Agency (1997b), National Ambient Air Quality Standards for Particulate Matter, *Fed. Regist.*, 62 (July 18, 1997) p. 138.

Vekeny, H. (1971), Gravimetric determination of the respirable dust fraction with the aid of a sampling device with an air sifter-pre-separator, *Staub Reinhalt. Luft*, 31, pp. 16-22.

Vincent, J.H., Aitken, R.J. and Mark, D. (1993), Porous plastic foam filtration media: their penetration characteristics and applications in particle size-selective sampling, *J. Aerosol Sci.*, 24, pp. 929-944.

Walton, W.H. (1954), Theory and size classification of airborne dust clouds by elutriation, *Br. J. Appl. Phys.*, 5 (Suppl. 3), pp. S29-S40.

Wesolowski, J.J., John, W., Devor, W. *et al.* (1977), Collection surfaces of cascade impactors, In: *X-Ray Fluorescence Analysis of Environmental Samplers* (T.G. Dzubay, Ed.), Ann Arbor Science Publishers, Ann Arbor, MI, pp. 121-131.

Wright, B.M. (1954), A size-selecting sampler for airborne dust, *Brit. J. Ind. Med.*, 11, p. 284.

Chapter 9

SUMMARY OF THE ADOPTED RECOMMENDATIONS

Morton Lippmann

Nelson Institute of Environmental Medicine,
New York University School of Medicine

9.1 Introduction to Particle Size-Selective Aerosol Sampling

There are a variety of reasons for determining the concentration and/or the size-distribution of airborne particles. Such data can be used to evaluate the nature and strength of sources of particles, as well as whether the concentrations exceed established air quality standards or guidelines. The various air quality criteria that have been developed are generally effect-related, such as for: a) product quality assurance (e.g., clean room applications); b) maintenance of visual range (for airport and highway safety and for esthetic considerations); or c) determinations of whether health-related concentration limits are exceeded. Some effects are caused by particles within limited size ranges and may be more closely related to the number of particles per unit volume than to their cumulative mass concentration, while others are more closely related to cumulative mass or at least mass within a specific range of particle sizes. In this document, the focus is on the scientific basis for collecting samples within specific particle size ranges, and especially within that range of particle sizes having the greatest influence on producing adverse health effects after deposition in the human respiratory tract.

Health effects associated with the inhalation of particles can be classified in terms of physiological, biochemical or anatomical changes in the airways of the respiratory tract or in other organ systems after the particles or their components are translocated to sensitive receptor sites. The effects can range from transient and reversible changes that produce discomfort or disability in the short-term to pathological changes associated with more permanent or progressive disease processes with fatal consequences.

Particle size can be a critical determinant of health effects in several ways. Firstly, while the particle is still airborne in the respiratory tract, particle size affects the extent and pattern of particle deposition on airway surfaces because of the size-dependence of the gravitational, inertial, diffusional, and electrical mobility phenomena which govern their behavior. The deposition patterns are also influenced by variations in airway dimensions and air flow (i.e., mean velocity, velocity profile and turbulence) that occur with depth into the lungs (see Chapter 2). Secondly, once deposited, particle size has a bearing on the probability of its remaining on the airway surface, where particle clearance to the gastrointestinal tract may be relatively rapid, or of migrating to the underlying epithelium. Particles adjacent to cell surfaces can be ingested by phagocytic cells. Once inside cells, retention can be prolonged. Alternatively particles can be dissolved, and their components made available for distribution to other organs by lymph and circulating blood. For very small insoluble particles (i.e., $\leq$ 30 nm) there can be rapid penetration through the epithelial lining of the airways into the interstitial spaces where they can cause acute inflammation. The influence of the airway fluids, and of the cellular types and functions, on particle clearance and retention are reviewed in greater detail in Chapter 2.

The chemical composition of the airborne particles also influences the likelihood and extent of adverse health effects because of its effects on solubility and metabolic processes that mediate translocation and biological responses. These influences are beyond the scope of this document. Further information is available in the ACGIH publication, *Documentation of Threshold Limit Values and Biological Exposure Indices* (ACGIH, 1993).

Health-based occupational and environmental exposure limits are generally based on responses observed in human volunteers or laboratory animals and, for aerosols, are increasingly reliant on knowledge of effects expressed for particles in specific size fractions. With this in mind, the balance of this document is specifically focused on health related aerosol sampling, primarily through the use of particle size-selective sampling as a means of restricting the analysis to that fraction of the overall aerosol that is most closely related to the risk of producing adverse health effects.

9.2 BACKGROUND

The initial primary charge to the ACGIH Air Sampling Procedures (ASP) Committee in 1981 was

> *". . . . to recommend size-selective aerosol sampling procedures which will permit reliable collection of aerosol fractions which can be expected to be available for deposition in the various major subregions of the human respiratory tract, e.g., the head, the tracheobronchial region, and the alveolar (pulmonary) region."*

It was recognized from the outset that, based on the physics governing the primary mechanisms of particle transport into and within the respiratory tract, the most relevant index of particle size was *particle aerodynamic diameter* (d_{ae}). Physically, this quantity is defined as the diameter of a sphere of density 10^3 kg/m^3 (the density of pure water) which has the same settling speed in still air under gravity as the particle of interest. As such, it therefore embodies not only the geometrical dimensions of the particle in question, but also its density and shape.

It was anticipated that the work of the Air Sampling Procedures Committee would lead to an approach by which the the ACGIH Chemical Substances Threshold Limit Values (TLV) Committee might establish particle size-selective threshold limit values (PSS-TLVs) for airborne agents occurring as aerosols. Such PSS-TLVs are a step toward improving validity, reliability, and precision in understanding risks, and therefore can lead to more appropriate protection of workers in many instances. Since the 1985 publication by the Air Sampling Procedures Committee of its first report (Phalen, 1985), there have been significant advances in the application of particle size-selective sampling approaches to the evaluation of inhalation hazards and toward the setting of improved standards. In summary, these are:

(i) In 1989, Soderholm, then Chair of the Air Sampling Procedures Committee, with the endorsement of the full committee, proposed modified particle size-selective sampling criteria for adoption by ACGIH, the International Standards Organization (ISO), and the Comité Européen de Normalisation (CEN), with the objective of achieving international harmonization (Soderholm, 1989). For respirable aerosol, representing the fraction of inhaled particles which may penetrate down to the alveolar region of the lung, the revised criteria effectively split the difference between the criterion adopted by ACGIH in 1968 and the criterion adopted by ISO in 1983 (that had been based on the original British Medical Research Council criterion first proposed as long ago as 1952). For thoracic aerosol, representing the fraction of inhaled particles that may penetrate into the lung below the larynx, the new criterion remained conservative in relation to the best available human tracheobronchial deposition data. For inhalable dust, representing the fraction of airborne particles that may enter through the nose and/or mouth during breathing, the criterion proposed in the original 1985 ACGIH report was retained. This initiative was well received by all the interested parties, and has been widely adopted internationally (ISO, 1992; CEN, 1992; ACGIH, 1999).

(ii) A 2 L/min impactor-based personal size-selective sampler with a sharp cut-off at d_{ae} = 0.8 µm was developed (Rubow *et al.*, 1990) for use in underground coal mines to separate diesel exhaust from coal mine dust. This was carried out on the basis of knowledge that those two aerosols had quite different particle size distributions; almost all of diesel exhaust particles are smaller then 0.8 µm while almost all of the coal mine dust particles are larger.

(iii) The United States Environmental Protection Agency (EPA) reviewed the basis for its National Ambient Air Quality Standard (NAAQS) for particulate matter (PM), and in 1997 promulgated a new fine particle standard to supplement its current thoracic dust standard (i.e., PM_{10}, which has an 50% 'cut' at d_{ae} = 10 µm) (EPA, 1996a and b). In principle, the new fine particle standard is based on a sharp 'cut' at d_{ae} = 2.5 µm on the basis of the different chemical constituents of the particles above and below that 'cut' size. The smaller particles, which are often most closely associated with excess mortality and morbidity in the community, are predominantly derived from combustion effluents, while the larger particles are largely dusts from mechanical processes and windborne soil.

In the above, while the diesel and ambient fine particle samplers may be regarded as "particle size-selective" and are focused on inhalation hazard evaluations, they differ in both 'cut-sizes' and 'sharpness-of-cut' from the more traditional particle size-selective aerosol samplers that are intended to mimic the more widely dispersed physical particle size-selectivity of the major functional components of the human respiratory tract.

This book is a revised and expanded version of the original 1985 ACGIH report, providing detailed reviews and discussions of the latest particle size-selective criteria and their applications in particle size-selective air samplers. Its coverage is extended to include applications to community air exposure assessments as well as occupational exposure assessments for airborne chemical agents.

9.3 Approach

In developing a rational approach toward the development of PSS-TLVs, the Committee took into account previous efforts at developing a framework of particle size-selective sampling for protection of humans, as well as the particle collection characteristics of the respiratory tract and the performance characteristics of modern sampling instruments. In addition, the Committee identified a workable procedure for the development of PSS-TLVs.

9.4 Applicability

Most of the material which follows on particle size-selective samplers that mimic the particle aerodynamic penetration characteristics of major component regions of the human adult respiratory tract is specifically intended for application to healthy adults. So it cannot be applied directly to more heterogeneous populations which, for example, may also include the seriously ill, the very young, or even the aged. Furthermore, certain circumstances arise even in the workplace that will result in loss of applicability of the Committee's recommendations. These circumstances may include high wind velocities or unusual wind patterns; high breathing rates (e.g., as occur upon heavy exertion); and unusual aerosols such as those which are hygroscopic, heavily electrically charged, or those which do not deposit within the respiratory system solely on the basis of their aerodynamic diameters (e.g., extremely long-thin fibers which deposit by interception).

It is important to note that the three sampling efficiency curves used to describe the inhalable, thoracic and respirable fractions respectively **cannot** be subtracted from one another in order to define deposition within a given anatomic region. For example, although the difference between the criteria representing the inhalable and thoracic fractions might idealistically appear to provide the corresponding criterion for deposition in the head airways region, this is not actually permitted. The reason for this apparent inconsistency is that, due to the uncertainty of the original data upon which the criteria were based, the penetration curves in question were chosen to err on the side of over-representing the true respiratory tract exposures, thus providing a degree of conservatism in the protection of workers when PSS-TLVs are proposed based on such fractions. This is especially the case for the thoracic fraction.

Further, because the respiratory tract penetration curves cannot be precisely matched by practical size-selective inlets and samplers, two different instruments, which both apparently conform adequately to the recommended curve, may not be in agreement in terms of the actual mass sampled. The degree of agreement may depend upon the characteristics of the sampled aerosol, in particular its particle size distribution. Therefore, characterization of aerosols for purposes other than establishing compliance with PSS-TLVs will, in general, require the use of additional techniques.

In terms of ambient air sampling for PM, the EPA PM_{10} standards for daily maximum and annual average concentrations for ambient atmospheric aerosol mandate the use of sampler inlets following characteristics that are essentially the same as those specified here for sampling the thoracic fraction. Thus, samplers designed for PM_{10} sampling may be useful for thoracic aerosol sampling and vice versa. By contrast, however, $PM_{2.5}$ samplers, with sharp 'cuts' at $d_{ae} = 2.5$ µm, developed for evaluating compliance with the July 1997 EPA fine particle NAAQS, will not be useful for evaluating occupational exposures to pneumoconiosis-producing dusts, nor — conversely — will respirable aerosol dust samplers, with a broad penetration curve around a 50% cut at $d_{ae} = 4$ µm, be very useful for the characterization of fine ambient particle mass concentrations. With this in mind, the discussion of $PM_{2.5}$ sampling in this book will be limited to the performance of samplers in meeting the EPA specifications.

9.5 Latest ACGIH Recommendations for Workplace Particle Size-Selective Sampling

As described in the main body of this book, the ACGIH Air Sampling Procedures Committee has considered many aspects of size-selective sampling in evaluating inhalation hazards in the workplace, including: a) effects of particle size on deposition site within the respiratory tract; b) the tendency for many occupational diseases to be associated with material deposited in particular regions of the respiratory tract; c) the availability of suitable samplers; and d) the relative inappropriateness of poorly defined "total dust" samples presently collected. As a result, the recommendations of the Committee may be summarized as follows:

(i) The Chemical Substances TLV Committee should continue to develop particle size-selective TLVs (PSS-TLVs), where "PSS-TLV" is a general term for:

a. *inhalable particulate matter TLVs* (IPM-TLVs) for those materials which are hazardous when deposited anywhere in the respiratory tract.

b. *thoracic particulate matter TLVs* (TPM-TLVs) for those materials which are hazardous when deposited anywhere within the lung airways and the gas exchange region.

c. *respirable particulate matter TLVs* (RPM-TLVs) for those materials which are hazardous only when deposited in the gas exchange region.

(ii) The size-selective sampling criteria should be based upon the following definitions:

a. The *inhalable particulate matter* fraction (IPM) consists of those particles that are captured according to the following collection efficiency averaged uniformly over all sampler orientations with respect to the wind direction:

$$\mathrm{IPM}(d_{ae}) = 0.5 \bullet \{1 + \exp(-0.06\, d_{ae})\} \qquad \textbf{9.1}$$

at least for $0 < d_{ae} \leq 100$ μm. Although, as described in Chapter 3, data are now starting to emerge for larger particles beyond this range, and for very low windspeeds not previously investigated, such new information does not at this stage provide a justification for any further adjustment of the above definition.

b. The *thoracic particulate matter* fraction (TPM) consists of those particles that are captured according to the following collection efficiency:

$$\mathrm{TPM}(d_{ae}) = \mathrm{IPM}(d_{ae}) \bullet [1 - F(x)] \qquad \textbf{9.2}$$

where $F(x)$ is the cumulative probability function of a standardized normal variable (x) given by

$$x = \ln\,(d_{ae}/\Gamma)\ /\ \ln\,(\Sigma) \qquad \textbf{9.3}$$

TABLE 9.1. Inhalable, thoracic, and respirable fractions

Particle Aerodynamic Diameter (μm)	Fraction Collected (%)
Inhalable	
0	100
1	97
2	94
5	87
10	77
20	65
30	58
40	54.5
50	52.5
100	50
Thoracic	
0	100
2	94
4	89
6	80.5
8	67
10	50
12	35
14	23
16	15
18	9.5
20	6
25	2
Respirable	
0	100
1	97
2	91
3	74
4	50
5	30
6	17
7	9
8	5
10	1

in which $\Gamma = 11.64$ µm and $\Sigma = 1.5$. The net result is that the function TPM reaches its 50% 'cut' at $d_{ae} = 10$ µm.

Here it is acknowledged that this mathematical form is unwieldy and so not convenient for routine use. With this in mind, the following simple analytical form provides results very close to the formal definition:

$$\mathrm{TPM}(d_{ae}) = \mathrm{IPM}(d_{ae})\left\{1 - \frac{\exp\{a + b\ln(d_{ae})\}}{1 + \exp\{a + b\ln(d_{ae})\}}\right\} \qquad \mathbf{9.4}$$

where

$$b = 1.658/\ln(\Sigma) \text{ and } a = -b \bullet \ln(\Gamma)$$

c. The *respirable particulate matter* fraction (RPM) consists of those particles that are captured according to the following collection efficiency:

$$\mathrm{RPM}(d_{ae}) = \mathrm{IPM}\ (d_{ae}) \bullet [1 - F(x)] \qquad \mathbf{9.5}$$

where $F(x)$ is again the cumulative probability function of a standardized normal variable (x) but with $\Gamma = 4.25$ µm and $\Sigma = 1.5$. The net result is that the function RPM reaches its 50% 'cut' at $d_{ae} = 4.0$ µm. The same simpler analytical formulae as that given above for the thoracic fraction may also be used.

These three fractions may also be given in tabular form, as shown in Table 9.1 (from ACGIH, 1999). The most significant difference from the definitions in the 1985 ACGIH report is the increase in the 50% 'cut' point for a respirable aerosol sampler from 3.5 µm to 4.0 µm, bringing it into accord with the ISO/CEN definition.

REFERENCES

American Conference of Governmental Industrial Hygienists (ACGIH) (1993), *Documentation of the Threshold Limit Values and Biological Exposure Indices, 6th Ed.*, ACGIH, Cincinnati, OH.

American Conference of Governmental Industrial Hygienists (ACGIH) (1999), *Threshold Limit Values and Biological Exposure Indices*, ACGIH, Cincinnati, OH.

Comité Européen Normalisation (CEN) (1992), *Size Fraction Definitions for Measurement of Airborne Particles in the Workplace*, CEN Standard EN 481, CEN, Brussels.

International Standards Organisation (ISO) (1992), *Air Quality - Particle Size Fraction Definitions for Health-Related Sampling*, CD 7708, ISO, Geneva.

Phalen, R.F. (Ed.) (1985), *Particle size-selective sampling in the workplace*, Report of the ACGIH Air Sampling Procedures Committee, American Conference of Governmental Industrial Hygienists, Cincinnati, OH.

Rubow, K.L., Marple, V.A., Tao, Y. and Liu, B.Y.H. (1990), Design and evaluation of a personal diesel aerosol sampler for underground coal mines, Preprint No. 90-132, Society for Mining, Metallurgy, and Exploration, Littleton, CO.

Soderholm, S.C. (1989), Proposed international conventions for particle size-selective sampling, *Ann. Occup. Hyg.*, 33, pp. 301-320.

United States Environmental Protection Agency (EPA) (1996a), *Air Quality Criteria for Particulate Matter*, EPA/600/P-95/001, Washington, DC.

United States Environmental Protection Agency (EPA) (1996b), *Review of the National Ambient Air Quality Standards for Particulate Matter*, Office of Air Quality Planning and Standards (OAQPS) Staff Paper, EPA-452/R-96-013, Environmental Protection Agency, Research Triangle Park, NC (July 1996).

Chapter 10

APPLICATION OF PARTICLE SIZE-SELECTIVE SAMPLING CRITERIA IN ESTABLISHING TLVs

Ronald S. Ratney

Mabbet and Associates, Bedford, MA

10.1 TLVs®: What are they?

Occupational exposure limits (OELs)

Over the years, many tables have been published of airborne concentrations of toxic substances that were considered safe. One of the earliest of these was published by Rudolf Kobert in 1912. The list provided information on 20 substances under four headings: 1) rapidly fatal to man and animals; 2) dangerous in one-half to one hour; 3) one-half to one hour without serious disturbances; and 4) only minimal symptoms observed after several hours. Sayers and Dallavalle (1935) later published a list of recommended exposure limits for 38 substances based on reports in the international toxicological literature. A series of historical papers on the development of occupational exposure limits in the United States was reprinted by the American Conference of Governmental Industrial Hygienists (ACGIH) in 1984 (LaNier, 1984).

The forerunner in the United States of the ACGIH "*threshold limit values*" (TLVs) and other occupational exposure limits (OELs) was a list of maximum concentrations assembled in 1937 from recommendations by the Massachusetts Division of Occupational Hygiene and faculty members at the Harvard School of Public Health and Yale

University (Bowditch *et al.*, 1940; Bowditch, 1944). Kobert's 1912 list does not identify exposures that were without effect but the text which accompanies the Massachusetts table indicates the authors felt that exposures below the recommendations would not have adverse health effects.

At the fifth annual meeting of the National Conference of Governmental Industrial Hygienists (the original name of the ACGIH) in 1942, a subcommittee on threshold limits presented a table of Maximum Permissible Concentrations of Atmospheric Contaminants (NCGIH, 1942). The list was a compilation of exposure limits used by various state industrial hygienists. In 1946 the subcommittee presented an expanded list (ACGIH, 1946) which included values from the 1942 report, a compilation by Professor Warren Cook (1945), and values published by the American Standards Association (now the American National Standards Institute). The list has been expanded and republished each year since 1946. In 1946 and 1947, the values were called "*maximum allowable concentrations*" but since 1948 they carried their current name, "*threshold limit values*" (TLVs).

The TLVs are occupational exposure limits (OELs) recommended by ACGIH, a not-for-profit professional society for occupational and environmental health and safety practitioners worldwide. It is not affiliated with any government agency. ACGIH publishes an annual list of TLVs in its *Threshold Limit Values and Biological Exposure Indices* book (ACGIH, 1999). In the preamble to the book, TLVs are defined as airborne concentrations of substances to which nearly all workers "*... may be repeatedly exposed day after day without adverse health effects.*" Adverse health effects include those that shorten life expectancy, compromise physiological function, impair the capability for resisting other toxic substances or disease processes, or adversely affect reproductive function or development processes. Some values are established to provide reasonable freedom from irritation, narcosis, nuisance, or other forms of stress.

Occupational exposure limits are established to protect normal healthy adults whose physical conditions permit them to work outside the home. The preamble to the ACGIH TLV/BEI book states that TLVs are designed to protect "*...nearly all workers.*" As discussed later in this chapter, ACGIH has not identified a specific level of risk subsumed under this term. But quantitative risk evaluations have determined that the risks associated with the TLVs cluster around 0.001. The U.S. Occupational Safety and Health Administration (OSHA) and other official agencies frequently use this as a guideline for establishing their own OELs. This guideline is frequently based on considerations of statistical confidence or engineering feasibility. It should be recognized that a small fraction of exposed workers may be adversely affected at exposures below the TLV or other OELs. Other exposure control procedures such as isolation, substitution, medical surveillance and personal protective equipment should be used in order to minimize exposure and its effects.

Occupational exposure limits refer only to exposures in the workplace and cannot be used as guides to exposures in the general environment or in residences. Persons outside the work environment include children, the elderly, and persons whose health may have been impaired by infections and other disease processes. Children undergo rapid growth and cell turnover and have undeveloped homeostatic defense systems. The elderly may have defenses which have been compromised by degenerative diseases or earlier toxic exposures. Also, some persons of working age may not be employed because of health conditions.

Another reason why OELs cannot be extended to the general environment is that workplace exposures normally occur only during the 2080 or so available working hours in a year, with detoxification and elimination occurring during nights, holidays and weekends when there are no exposures. Environmental exposures can continue during these times, which would provide an increased period for the intake of toxic substances with little opportunity for detoxification and excretion.

Occupational exposure limits are published by governmental and non-governmental bodies around the world. In several countries, the OELs are legally binding and have the force of law. OELs published by non-governmental bodies and by some governments are professional recommendations developed for the guidance of industrial hygienists and plant managers. For many years, government agencies and professional organizations adopted the then current TLVs. But more recently OELs have been developed independently through a review of the scientific and medical literature. Vincent (1998) has recently reviewed the establishment of OELs in Australia, Britain, Norway, and Russia based on personal discussions with occupational health scientists and officials in those countries. Australia, Britain, and Norway initially used the TLVs as their national limits but have recently moved towards developing OELs independently. The Russian OELs are unique in that they are based almost exclusively on toxicological studies conducted in Russia and reported in Russian journals and monographs. Most of these sources are not widely available outside Russia. However, the United Nations Environmental Program published an English translation of a compendium of Russian OELs for workplace air, residential areas, and drinking water, which contains very brief descriptions of the toxicological data supporting the recommendations (Izmerov *et al.*, 1982).

The International Labour Office (ILO) published a compilation of OELs from 15 countries in 1991. The countries represented in the list were Australia, Belgium, Czechoslovakia, Denmark, Finland, France, Germany, Hungary, Japan, Poland, Sweden, Switzerland, United Kingdom, United States, and the former Soviet Union. In Germany, the OELs are established by an independent professional organization, the Deutsche Forschungsgemeinschaft; in Japan OELs are developed by the Japan Association of Industrial Health. In the U.S., OSHA legally enforces a set of "*permissible exposure limits*" (in the form of its annually published PELs), which it promulgates after a review of the medical and scientific literature and participation by industry and organized labor. The National Institute for Occupational Safety and Health, a unit of the U.S. Centers for Disease Control and Prevention (CDC), develops "*recommended exposure limits*" (RELs) as professional recommendations (NIOSH, 1997). ACGIH, a professional association with no organizational ties to the U.S. government, annually publishes an updated list of TLVs as professional recommendations.

The Health and Safety Directorate of Directorate General Five (DGV) of the Commission of the European Communities (CEC) has promulgated OELs for lead, asbestos, vinyl chloride, and a group of about 50 substances in its 1978, 1982, 1990, 1991, and 1996 directives. It will be promulgating additional limits in the future. However, the OELs developed by DGV do not automatically become legally binding limits in the countries of the European Union (EU). Members of the EU are required to adopt their own laws and regulations which conform — at a minimum — to the provisions of the relevant DGV directives.

The most widely used OELs have been published as professional guidelines for industrial hygienists, plant managers and occupational health professionals. ACGIH has already been mentioned and the procedures of its Committee on TLVs for Chemical Substances are described later. A similar organization in Germany is the Commission for the Investigation of Health Hazards of Chemical Compounds in the Workplace which is a unit of the Deutsche Forschungsgemeinschaft. This group publishes annually an extensive list of "*maximum arbeitsplatz koncentrations*" (MAKs) and "*biologischer arbeitsstofftoleranz-werts*" (BATs) with accompanying documentation (MAK, 1997). In the United States, the American Industrial Hygiene Association (AIHA, 1996) has an ongoing program for the development of "*workplace environmental exposure levels*" (WEELs) and "*emergency response planning guidelines*" (ERPGs). Most of the substances covered by WEELs do not have TLVs but there are a few overlaps. Elsewhere, in 1995, a self-constituted health-based exposure limits committee of the Santa Clara Center for Occupational Safety and Health published a list of OELs which are generally lower by several orders of magnitude than values recommended by other organizations.

In many countries, government agencies promote worker health by promulgating legally binding OELs. Depending on the laws of the particular country, these may be treated as absolute limits not to be exceeded on any single day. Government inspectors will evaluate exposures in a workplace, and if overexposures are found, the employer may be subject to legal sanctions varying from monetary fines through plant closure. In many countries, existing professional recommendations such as the TLVs are adopted as legal limits but in the U.S. and several countries in Europe, limits are developed independently through a formal legal process. In all cases, there are critical evaluations of the scientific and medical literature on the substances being considered, but feasibility, economic impact, and the views of business and labor are incorporated into the determinations. Legal OELs are usually incorporated into more inclusive laws and regulations that encompass protective measures besides control of airborne exposures, including medical surveillance, personal protective equipment, work practices, and routine exposure monitoring.

The ACGIH Chemical Substances TLV Committee

The Committee on Threshold Limit Values for Chemical Substances is one of the technical committees of ACGIH. Full membership in ACGIH is open to industrial hygienists, occupational health, environmental health or safety professionals whose full-time employment is with a governmental agency or educational institution and who are engaged in health or safety services, standard setting, enforcement, research or education, including professionals outside the United States who meet the employment requirements. Affiliate membership is available to qualified professionals who are not employed by governmental agencies or educational institutions. This includes employees of corporations, labor unions, independent research institutes, trade associations, and private consultants.

All members of the TLV Committee are also members of ACGIH. Affiliate and honorary members of ACGIH, who may be employees of corporations, labor unions and other non-governmental organizations or educational institutions, may participate fully in the deliberations of the Committee but may not vote on official recommendations. The members of the Committee are selected on the basis of their experience and expertise and not as representatives of their employers. Because the development of a TLV requires the integration of data on toxicology, industrial hygiene, and occupational medicine, there is a deliberate effort to select persons with expertise in these disciplines for membership on the Committee.

On occasion, the full membership of the German MAK Commission meets with the Chemical Substances TLV Committee and participates actively in its deliberations. In addition, a member of the AIHA WEEL Committee attends all meetings of the Chemical Substances TLV Committee to provide for liaison with that organization.

The Chemical Substances TLV Committee has four standing subcommittees. The first three are: 1) HOC, covering organic compounds of carbon, hydrogen and oxygen; 2) MISCO, covering organic compounds containing chlorine, nitrogen and other elements; and 3) Dusts and Inorganics, covering inorganic gases and materials which occur as airborne particles. The fourth subcommittee, on carcinogens, provides guidance to the subcommittees and the full committee on the classification of substances as carcinogens.

When available information indicates that a new substance should be added to the TLV list, or that an existing TLV should be changed, the chair of the appropriate subcommittee assigns a member the task of reviewing the scientific literature, preparing a draft documentation, and recommending a TLV. The subcommittee members critically review the draft, the significant original papers, and the recommended value. The author may be asked to make revisions and the review cycle may continue at another meeting. Finally, after intense discussion, a consensus position is reached and the subcommittee transmits its recommendation to the full Committee for further action.

The full Committee reviews the subcommittee's recommendation, frequently with new insights provided by the Committee members. It is not uncommon for proposals to be sent back to a subcommittee for further revisions. When the Committee agrees that a proposal is well-supported by the information presented in the documentation, it will vote formally to transmit the recommendation to the ACGIH Board of Directors. Any proposed new TLV, whether for a new substance or as a revision of an existing value, is initially published on a "Notice of Intended Changes" (NIC) to permit interested parties an opportunity to provide additional data and comments. Occasionally industry groups will meet with a subcommittee to present information and views.

The Chemical Substances TLV Committee meets semi-annually, in Spring and Fall. The Spring meeting is used to discuss general concepts as well as to review documentations and TLV proposals that have been circulated since the last meeting. In Fall, the full Committee receives recommendations from the subcommittees concerning proposed TLVs to be placed on the NIC list or adopted as final TLVs. After discussing the documentations circulated by the subcommittees, the full Committee votes to accept the recommendations, change them or send the issues back to the subcommittee. The decisions of the full Committee are transmitted to the ACGIH Board of Directors. When the Board has ratified the recommendations, they are published in "Today!," the ACGIH newsletter, and on the ACGIH website, www.ACGIH.org. The recommendations are also available by request to ACGIH headquarters and are reprinted with permission in other publications covering industrial hygiene and occupational safety.

10.2 Pathological conditions caused by deposited particles

As described in Chapter 2, the various anatomic structures of the respiratory tract (e.g. the nose, trachea, bronchi and alveoli) have been allocated to one of three respiratory tract regions: the head airways region (HAR); tracheobronchial region (TBR); and gas exchange region (GER). The penetration of particles into one of these regions is a function of particle size as discussed in Chapter 1. Particles are also cleared from their sites of deposition by mechanisms and at rates that are characteristic of the site of deposition. Accordingly, after inhaled particles have deposited on a surface of the respiratory tract, they may express their toxicity at the site of deposition, or they may be translocated to other parts of the body to express their toxicity elsewhere. Deposited particles may: (a) dissolve in pulmonary fluids and pass into the systemic circulation; (b) pass directly through the alveolar walls into the interstitial spaces, capillaries or the lymphatic circulation; (c) move into the throat through the action of the "ciliary escalator"; (d) be captured by alveolar macrophages which then move into the ciliated regions of the lung; or (e) bind to local tissue.

When particles initiate a pathological condition, the sites of deposition are critical parameters in the establishment of a TLV based on particle size. The first step in the development of a particle size-selective TLV is therefore the identification of the disease that is associated with particle inhalation and the anatomical location where the pathological process is initiated.

Diseases are usually diagnosed on the basis of symptoms reported by patients or clinical tests. The anatomical locations of pathological lesions can frequently be determined on the basis of signs, symptoms, and tests, but in some cases, only biopsy observations can pinpoint the sites. Clinical diagnoses are usually coded in hospital records and on death certificates in accordance with the fifth edition of the International Classification of Diseases (ICD), 9th Revision (United States Department of Health and Human Services, 1994). Table 10.1 lists the ICD codes for diseases of the respiratory tract and identifies the corresponding regions defined in Chapter 2. The classifications identify the general anatomical regions of the respiratory tract associated with the listed diagnosis but these do not always correspond to the anatomical regions defined in Chapter 2 in terms of particle deposition. Non-infectious diseases of the upper respiratory tract (classifications 470–478) include conditions of the nose, pharynges, nasopharynges, sinuses, and larynx. These are the same anatomical regions identified as the HAR. Classifications 490–496 are described as "*chronic obstructive pulmonary disease and allied conditions*" (COPD). The diseases included here are acute and chronic bronchitis, emphysema, asthma, bronchiectasis and alveolitis, all of which involve both the thoracic and gas exchange regions. Most lung diseases due to inhaled particles are included in classifications 500–508, "*pneumoconioses and other lung diseases due to external agents*," but conditions in the thoracic and gas exchange regions are sometimes grouped together. Classification 506.4, "*chronic respiratory conditions due to fumes and vapors*" includes emphysema, bronchiolitis, and fibrosis. However, allergic alveolitis (495) and extrinsic asthma (494) are included with COPD. Classifications 500 through 508 are used when inhaled dust has been identified as the etiological agent for a condition. If a physician has not identified a dust as the cause of the observed signs and symptoms, the condition may be coded under one of the more general classifications. Table 10.1 identifies the respiratory tract regions, as used in this monograph, associated with specific diagnostic categories.

In the simplest analysis, it is assumed that when a sufficient quantity of toxic particles deposits on a respiratory tract surface, it may initiate a disease process on that surface. This will occur when the particles become fixed at the site of deposition and the tissues initiate a toxic or foreign body response. Particles can also elicit chemotactic agents

TABLE 10.1. International Classification of Diseases associated with specific regions of the respiratory tract

ICD Code	ICD Name	Notes	Respiratory Region*
460.0	Acute rhinitis	Usually the common cold	HAR
472.0	Chronic rhinitis		HAR
472.1	Chronic pharyngitis		HAR
472.2	Chronic nasopharyngitis		HAR
473.0	Chronic sinusitis		HAR
476.0	Chronic laryngitis		HAR
476.1	Chronic laryngotracheitis		HAR
477.0	Allergic rhinitis		HAR
478.0	Other diseases of upper respiratory tract (includes nasal cavity, pharynx, larynx and trachea)		HAR (except trachea which is TBR)
490.0	Bronchitis, not specified as acute or chronic	Not applicable. Bronchitis due to fumes and vapors are included in 506.0	
491.0	Chronic bronchitis	May be coded as 506.0	TBR
492.0	Emphysema		GER
493.0	Asthma		TBR
494.0	Bronchiectasis		TBR
495.0	Extrinisic allergic alveolitis		GER
496.0	Chronic airway obstruction, not elsewhere classified		TBR
500.0	Coal workers' pneumoconiosis		GER
501.0	Asbestosis		GER
502.0	Pneumoconiosis due to other silica or silicates		GER
503.0	Pneumoconiosis due to other inorganic dust		GER
504.0	Pneumopathy due to inhalation of other dust	e.g., Byssinosis	
506.0	Chemical bronchitis		TBR
506.1	Chemical pulmonary edema		GER
506.2	Upper respiratory inflammation due to fumes and vapors		HAR
506.4	Chronic respiratory conditions due to fumes and vapors	Emphysema, Obliterative bronchiolitis, Pulmonary fibrosis	GER
507.8	Detergent asthma		TBR

*TBR = tracheobronchial region
GER = gas exchange region, or alveolar region
HAR = head airways region

or other forms of "*biochemical lesions*" which initiate pathological processes elsewhere in the body after the particles have been cleared from the pulmonary tract.

Very, very small particles deposited in the lung may move through the alveolar walls and enter the lymphatic or circulatory system or become sequestered in the tissues surrounding the lungs, such as the pleura or mesothelium. Particles that reach the alveoli can be captured by alveolar macrophages which migrate into the bronchioles and other proximal parts of the pulmonary system where disease processes can develop remote from the original site of deposition. Finally, particles may dissolve in pulmonary fluids through physical–chemical reactions, or by conjugation with proteins, and then enter the systemic circulation.

Bioaerosols, which include bacteria, viruses, spores, and other infectious or antigenic proteins, usually act through an immunological response that can be remote from the site of deposition. Microorganisms such as bacteria, many of which have particle aerodynamic diameters of 1–2 µm, and viruses with diameters of 0.01–1 µm, may penetrate the alveolar parenchyma, remain at the site of deposition or elicit antibodies which then enter the systemic circulation. Fungal spores have diameters of 3–30 µm and most of those inhaled are captured largely in the HAR. A common cause of allergic rhinitis, they also can cause asthma and hives through an immunological response. Finally, non-living antigenic proteins such as subtilisins can cause pulmonary symptoms through immunologic mechanisms.

Because of the circumstances just discussed, it is necessary to elucidate the mechanism through which inhaled particles elicit an adverse health effect. If a disease occurs in the GER, the TBR or HAR, one should not necessarily assume that the offending particles are those residing in the respirable, thoracic or inhalable fractions.

10.3 Dose metrics

The establishment of a TLV essentially requires the determination of a relationship between the dose (exposure) to a toxic substance and the associated response (adverse health effect).* At each dose, some fraction of the exposed humans or animals will be affected or

*Here, for the purposes of this discussion, these "macroscopic" terms are used, although it is acknowledged that they may have more specific meanings to individual researchers. For example, "dose" might be some quantity involving the amount of the substance present in relevant tissue and the time it spends there, while "response" might refer to the initial local cellular response arising from the dose there.

there will be an alteration in one or more physiological parameters. In many investigations, there will be a dose or exposure at which no adverse effect is observed and, as discussed later, the TLV will be set at, or below, this "*no observed adverse effect level*" (NOAEL).

In animal toxicology experiments where toxicants are administered percutaneously, the doses are carefully specified and changes in physiological parameters are measured, or the incidences of a specified outcome such as death or cancer are observed. Where animals are dosed by the airborne route and in occupational epidemiology studies, the dose of toxicant actually administered is not measured directly. Rather, the toxicant concentration in the air available for the subject to breathe is measured. It is assumed that the amount of material transferred to the respiratory tract, and then to the rest of the body, is an unknown fraction of the amount of substance in the air, which depends on the particle size, breathing rate, and other characteristics of the subject and the environment. This is particularly true in epidemiology studies where airborne concentrations in the general vicinity of the workers may be the only available measure of administered dose.

The vast majority of workplace exposures in most industries have been measured traditionally in terms of 'total' dust, using that as a surrogate for the dose actually entering a worker's body. In the early days of industrial hygiene, exposures were measured primarily to track the effectiveness of ventilation and other control measures. It was recognized that decreased exposure resulted in decreased risk of disease, but only in a very qualitative sense. In addition, air sampling instruments were large and cumbersome and ill-adapted for measuring personal exposures or for collecting particles in different size ranges.

Most early reported exposure measurements do not specify the sampling instruments used but it can generally be assumed that they were impingers, konimeters or devices aimed at providing samples suitable for assessing exposures in terms of particle counts, or they were high-volume samplers used for collecting area samples which were then assessed gravimetrically. Such methods have been reviewed in a recent essay by Walton and Vincent (1998). Different types of samplers collect different fractions of the airborne particle load and rarely mimic the collection efficiency of the human respiratory system. Nevertheless, in the absence of definitive data, it is usually conservatively assumed that the reported contaminant concentration is the same as that which enters the human respiratory tract (or, in some cases, penetrates to the alveolar region of the lung).

Because personal sampling instruments were not widely available until the 1970s, most earlier measurements were taken in the general plant air or in the vicinity of the workers. Air contaminant concentrations in such area samples bear only a qualitative relationship to exposures measured in a worker's breathing zone. It has been widely observed that contaminant concentrations in area samples can be substantially higher or lower than worker breathing zone measurements, depending on the relative positions of the worker and the sampler with respect to the source of contamination and exhaust ventilation.

The use of "*respirable mass*" as the dose metric for TLVs for dusts which affect the gas exchange regions of the lung has been recognized since 1968. At that time, ACGIH proposed TLVs for respirable dusts containing quartz, cristobalite, and tridymite (ACGIH 1968). Equipment for measuring respirable particulate has been available for many years and exposures to fibrogenic dusts are usually reported in terms of the "*respirable mass concentration.*"

Although it has been recognized that mass of respirable particulate was the appropriate dose metric for fibrogenic dusts, there were several different definitions of the specific size fractions to be included. The differences between these definitions are discussed elsewhere in this publication. Instruments were available for measuring airborne dust concentrations in accordance with the different definitions. Respirable particulate defined in accordance with the original ACGIH recommendation had a 50% cut point at d_{ae}=3.5 µm. A revised respirable mass criterion adopted by the ACGIH in 1993 has a 50% cut point at d_{ae} = 4.0 µm (ACGIH, 1993a). In the U.S. and many other countries, cyclone pre-separators are used which were thought to conform to the original 3.5 µm cut criterion. Recent research has shown that these instruments actually conform more closely to the current 4.0 µm cut (Bartley *et al.*, 1994). The 50% cut point for the British Medical Research Council (BMRC) criterion, the earliest of all the particle size-selective criteria dating back to the earlier 1950s, is 5.0 µm (BMRC, 1952). So too is the Johannesburg convention, which derived directly from the BMRC criterion (Orenstein, 1960). As a result of the differences between the conventions, airborne dust measurements taken with a sampler conforming to the BMRC convention will be higher than samples taken in the same environment with a pre-separator conforming to the ACGIH definition. The ratio between the two measurements will depend on the particle size distribution of the airborne dust.

As indicated, reported respirable dust concentrations taken with different instruments may not be exactly comparable. The differences, however, are likely to be far less than the variability of environmental measurements or the uncertainties produced by using a few measurements taken in a small number of surveys to represent exposures of a large number of workers over a long period of time. Accordingly, respirable dust concentrations reported by different investigators are usually incorporated directly into the determinations of TLVs without adjusting for the different samplers that may have been used (and so do not allow for the inherent systematic biases — which may be considerable). The adoption of the revised ACGIH criterion by the International Standards Organization (ISO, 1992) and the Comité Européen de Normalisation (CEN, 1992) should encourage the use of samplers conforming to that criterion so that respirable dust measurements reported in the future should be more comparable.

Substances that elicit adverse effects after deposition in the TBR of the respiratory tract are assigned TLVs for particles collected by a sampler conforming to the thoracic convention. At the time of this publication, however, no personal samplers for the thoracic fraction of airborne dust were generally available. Accordingly, no measurements of exposure to thoracic particulate fraction of inhaled contaminants have been published and, in turn, it has not been possible to develop TLVs which refer to this aerosol fraction. 'Total' aerosol measurements have been published for many substances that affect the TBR, and data are also available for health effects associated with measured total dust exposures. In these cases, TLVs are still currently expressed in terms of 'total' dust. As personal thoracic dust monitors become available, TLVs for this fraction will be developed.

Materials which express their toxic effects anywhere in the body, regardless of the region of the respiratory tract in which they deposit, should be sampled with a collector that samples the inhalable fraction. Samplers conforming to the inhalable convention are becoming more available and exposures expressed as the inhalable fraction are appearing in the literature. However, TLVs are developed using exposure measurements reported in the referenced literature sources. Usually, the types of sampling instruments that were used are not identified but the context of the reports implies that the reported dust concentrations represent "total particulate." Even now, most exposures are measured with 'total' aerosol samplers to determine compliance with government regulations. In the U.S. and many other countries, exposure limits for aerosols are still expressed in terms of 'total' dust because the limits were based on the TLVs extant at the time of promulgation, or similarly derived values. These limits in turn were derived from exposures measured as 'total' dust. In the U.S. there are as yet no formal regulations which require sampling for the inhalable fraction; most inhalable dust measurements are carried out as parts of research studies. In the United Kingdom, the Health and Safety Executive (HSE) now requires that airborne exposures formerly designated as "total dust" are to be monitored in terms of inhalable particulate matter (IPM). In fact, it has done so — in principle at least — since 1989. Similarly the German MAK Commission now expresses exposure limits formerly designated as 'total' aerosol in terms of IPM. Other countries are following suit. Such initiatives will increase the use of inhalable samplers and the publication of exposures to inhalable particulate and their resultant health effects.

Because only the inhalable fraction of airborne particulate can have an effect on the body, (i.e., particles that are not inhaled cannot be harmful), it is clear that current TLVs expressed in terms of 'total' should be replaced with inhalable TLVs. In 1994, with this in mind, ACGIH proposed that TLVs hitherto expressed in terms of 'total' aerosol would be specified in terms of inhalable particulate. However, side-by-side field measurements using the IOM inhalable sampler (see Chapter 6) and samplers collecting 'total' aerosol (e.g., the closed-face 37-mm plastic cassette) have shown that inhalable aerosol concentrations tend to exceed the corresponding 'total' aerosol concentrations almost everywhere, except in welding environments where the two measurements were found to be approximately equal (Werner *et al.*, 1996). In general, the ratio of inhalable to total dust ranged from 1.29 to 3.57. From this it becomes apparent that, if the inhalable designation were applied uniformly to all TLVs now expressed as 'total' particulate, but with no adjustment in the numerical values, exposure measurements obtained using inhalable aerosol samplers would be higher, even without any **actual** change in the aerosol concentration. As a result, for environments with exposures only slightly below the 'total' aerosol TLV, the new inhalable aerosol exposures may well rise above the TLV, sometimes by a considerable margin. So it is possible that a plant environment which is in compliance with a current 'total' aerosol TLV might be out of compliance with a new inhalable aerosol TLV, and in turn would require additional exposure controls in order to come back into compliance. Again, it is stressed, this situation arises even though nothing in the environment has **actually** changed. This situation is likely to cause consternation both to industrial hygienists and to industry. But it underlines sharply the importance of understanding what it is that is being measured and how that measurement relates to the aerosol-related health effects which are the primary object of concern.

Based on concerns about the potential disruptive effects of a switch away from the 'total' aerosol convention, ACGIH in 1995 rescinded the proposal of a year earlier and again stated that TLVs for particulates would continue to be expressed in terms of 'total' aerosol unless the respirable, thoracic or inhalable fraction was specifically noted in the TLV list and in the *Documentation of the Threshold Limit Values and Biological Exposure Indices* (ACGIH, 1993b). Any proposed changes in the particle size designation would first be published in the NICs to permit interested parties to comment on the changes.

Notwithstanding the above, it is clear that aerosol exposure measurements are surrogates for inhaled doses, which in turn are surrogates for the risk of disease. A TLV corresponds to an airborne exposure and inhaled dose at which no adverse health effects will occur. At the same time, a TLV is a benchmark for efforts to control risk. If a TLV is to serve as an accurate benchmark, it must be expressed in the same units as the airborne measurements against which it is compared. This therefore means that if industrial hygienists will be measuring exposures in terms of IPM, it is scientifically consistent that the TLVs must also be in terms of IPM.

As industrial hygienists increasingly use samplers conforming to the IPM criterion, it will be necessary to convert the 'total' aerosol TLVs to ones referred to IPM. But because the ratio between the two exposure metrics depends on the particle size distribution of the aerosol being sampled, which in turn is controlled by the aerosol-generating process in question, the ratio will not be a constant even for a single substance. Strictly speaking, therefore, the only accurate way of determining the factor for converting 'total' aerosol TLVs to ones based on IPM would be to measure airborne concentrations simultaneously with both types of samplers. However, a review of such comparisons that has been published indicates conversion ratios for specific processes occur in narrow ranges. With this in mind, Werner *et al.* (1996) published a list of conversion factors for commonly used industrial processes (see Table 10.2).

It is now inevitable that all TLVs for substances which occur as airborne particulate will eventually be expressed in terms of either IPM, thoracic particulate matter (TPM), or respirable particulate matter (RPM). So when the Chemical Substances TLV Committee develops new TLVs for inhalable particulate, it will need to incorporate published exposures expressed as 'total' aerosol into dose–response relationships with doses re-expressed in terms of IPM. In doing so, published 'total' aerosol exposures must be converted to inhalable particulate, for example, by using the conversion factors suggested in Table 10.2. The Chemical Substances TLV Committee is also considering ways of changing existing TLVs expressed in terms of 'total' aerosol to values expressed in terms of IPM. This issue is discussed further below.

In 1996, sixteen substances were assigned respirable TLVs. Modified TLVs for three substances were on the NIC with proposed respirable TLVs. Inhalable TLVs were identified for eight substances. Cadmium and diquat had both respirable and inhalable TLVs to recognize that those substances were associated with both systemic and pulmonary effects. The relevant particle fraction is not identified for the vast majority of substances which occur as airborne particles. By default, these TLVs are taken to be expressed in terms of 'total' aerosol (ACGIH, 1997). An industrial hygienist who wishes to evaluate an environment for a particular substance with a 'total' aerosol TLV, but using an inhalable sampler, might choose an appropriate conversion factor from Table 10.2 and use it to scale down the measured exposure. Alternatively, the 'total' aerosol TLV may be scaled **up** by the same factor and compared with the actual measurement.

In Germany, the MAK Commission changed all 'total' aerosol MAKs to inhalable with no changes in the numerical values. The United Kingdom HSE did the same with its enforceable OELs. In the U.K. the aerosol sampler recommended for inhalable aerosol since 1989 has been the so-called "*7-hole sampler*," under the assumption — based on earlier wind tunnel research — that this sampler provided a good match with IPM. So no conversions were considered necessary. However, recent field studies in Australia reported by Terry and Hewson (1996) have shown that the 7-hole sampler actually undersamples in relation to the IOM sampler widely considered as the best reference sampler for IPM.

Throughout this chapter, and indeed throughout this publication, it has been assumed that an appropriate dose metric is the mass concentration of airborne particles (e.g., in mg/m^3), be it for 'total', inhalable, thoracic or respirable particulate. Particularly in the context of what some have referred to as "*very, very small*" particles (i.e., those with diameters less than 0.1 μm), some workers have suggested that the **number** concentration of airborne particles, or even their **surface area** concentration, might provide a better correlation with migration of particles to the pulmonary interstitium and associated inflammatory responses (Oberdorster *et al.*, 1992; Donaldson *et al.*, 1999). However, such correlations have been observed in only a small

TABLE 10.2. Suggested working conversion factors (S- values) for use where it is deemed desirable to adjust exposure data to account for the change in exposure assessment rationale, based on generalization of results of comparisons between 'total' aerosol as measured using the closed-face 37-mm sampler and inhalable aerosol as measured using the IOM inhalable dust sampler (Werner *et al.*, 1996).

Aerosol Classification/Industrial Category	Suggested Conversion Factor
Dust —	2.5
Mining	
Ore and rock handling	
Handling/transporation of bulk aggregate	
Textiles	
Flour and grain handling	
etc.	
Mist —	2.0
Oil mist and other machining fluids	
Paint sprays	
Electroplating	
etc.	
Hot processes	1.5
Metal smelting and refining	
Foundries	
Welding—	1.0
All	
Smokes and fumes —	1.0
All	

number of animal experiments and it seems likely that, in the standards setting arena, particle mass concentration will remain the dose metric of choice until a larger body of animal and human exposure and health data has been accumulated.

10.4 Metrics for Adverse Effects

In animal experiments, the adverse effects (lethality) of chemicals are traditionally quantified in terms of the fraction of test animals dying at each dose. The measure of toxicity for airborne chemicals may be expressed as the "*lethal concentration-50%*" (LC_{50}), while those for chemicals administered in discrete doses can be expressed as the "*lethal dose-50%*" (LD_{50}), the dose at which half the animals are killed by the chemical. For human epidemiological studies, the corresponding measure of effect is the age-adjusted death rate for specific causes of death. As interest has shifted from lethal to non-lethal effects, both animal and human studies have examined morbidity as well as mortality. Effects studies include respiratory symptoms such as asthma and bronchitis, transient irritation, neurological effects (e.g., tremor), and reproductive effects (e.g., infertility).

In dose–response studies, a subject is identified as displaying or not displaying the effect in question. Where mortality is being studied, the endpoint is clear. But for non-lethal effects, the degree of effect must be considered. Standardized criteria exist for several endpoints. For respiratory symptoms, the questionnaire developed by the BMRC when administered by experienced investigators provides an appropriate level of standardization in identifying outcomes. However, such terms as "*abnormal lung function*" or "*pneumoconiosis*" must be specified in terms of standardized criteria. They may even have slightly different meanings in different countries.

Lung function is defined in terms of pulmonary function parameters such as forced expiratory volume in one second (FEV_1) or forced vital capacity (FVC), both of which are quantitative measures. The values chosen to represent abnormal function must be specified recognizing the clinical significance of the observed effects. Careful evaluation can identify decrements in FEV_1 or FVC of 10% or less, but these may not be disabling or even recognizable by a subject. In addition, statistically significant declines in pulmonary function can be observed over time. But here too, the clinical significance of the decline must be evaluated.

Pulmonary fibrosis is frequently evaluated in terms of categorization of radiographic abnormality using the ILO scheme. The ILO classification of individual radiographs can be identified with some precision by a panel of experienced radiologists. The ILO category corresponding to the term "pneumoconiosis" should be carefully specified.

Quantitative physiological parameters can also be used directly as effect metrics. These parameters may be direct measures of physiological functions such as FEV_1, enzyme activity or other measurable variables. Table 10.3 lists some commonly used parameters.

TABLE 10.3. Some physiological parameters associated with toxic substances

Parameter	Inhaled Toxicant
FEV_1	Silica
β 2-Microglobulin	Cadmium
Cholinesterase Activity	Organophosphate Pesticides
Hand Steadiness	Mercury, Manganese
Particle Clearance	Insoluble Particles

While it is possible to measure many physiological parameters quite precisely, the difference between the observed and normal values must be significant in terms of the clinical status of the affected person. This issue was discussed by Hatch in 1977 who separated the dose-effect continuum into no-effect, normal adjustment, impairment, disability and death. Hatch embodied this idea in a graph reproduced here as Figure 10.1, about which he said:

> *A distinction is made between impairment and disability, the two scales representing, respectively, the underlying disturbance of the system and the consequence of such disturbance in terms of identifiable disease. Starting with normal health, the individual progresses, for one reason or another along the scale of impairment and disability, ultimately to death. Early departures from health (impairment) are accompanied by little disability. In the beginning, the normal homeostatic processes insure adequate adjustment to offset stress and for a distance beyond this early zone of change, compensatory processes similarly maintain the overall function of the system without serious disability. Further increments in impairment beyond the limits of compensatory processes, however, are accompanied by rapidly increasing increments in disability and the individual moves into the region of sickness and disability, terminating in death. A healthy individual, functioning at point A on the curve and subjected to a given kind and degree of stress may respond with relatively minor and temporary disturbance and will return to his underlying position when the stress is removed. An individual at point B, on the other hand, may find the same kind and degree of stress intolerable and, in the consequence, move rapidly up the curve to a position of serious disability and even death. In our past concern with occupational diseases, relationships were established between conditions of exposure and degrees of disability and objectives were to bring the stresses of the job within limits to prevent such disability. For the future, concern must be with impairment, rather than with disability, and relationships have to be demonstrated between the stresses of the job and the more subtle disturbances. The degree of impairment must be kept within limits well below the level of disease.*

Although great strides have been made in occupational toxicology, Hatch's comments are still relevant today. His graph can be modified by plotting the value of a parameter affected by a toxicant. Under normal circumstances, the parameter varies with a range that corresponds to the homeostatic mechanisms of living systems. As long as a toxicant does not cause the parameter to move outside this range, no illness or disability occurs. Higher levels of the parameter will cause physiological changes that will be recognized as symptoms discernible by the subject or as frank clinical illness. However, after the exposure to the toxicant ceases, normal repair mechanisms will return the subject to a normal condition without future clinical consequences.

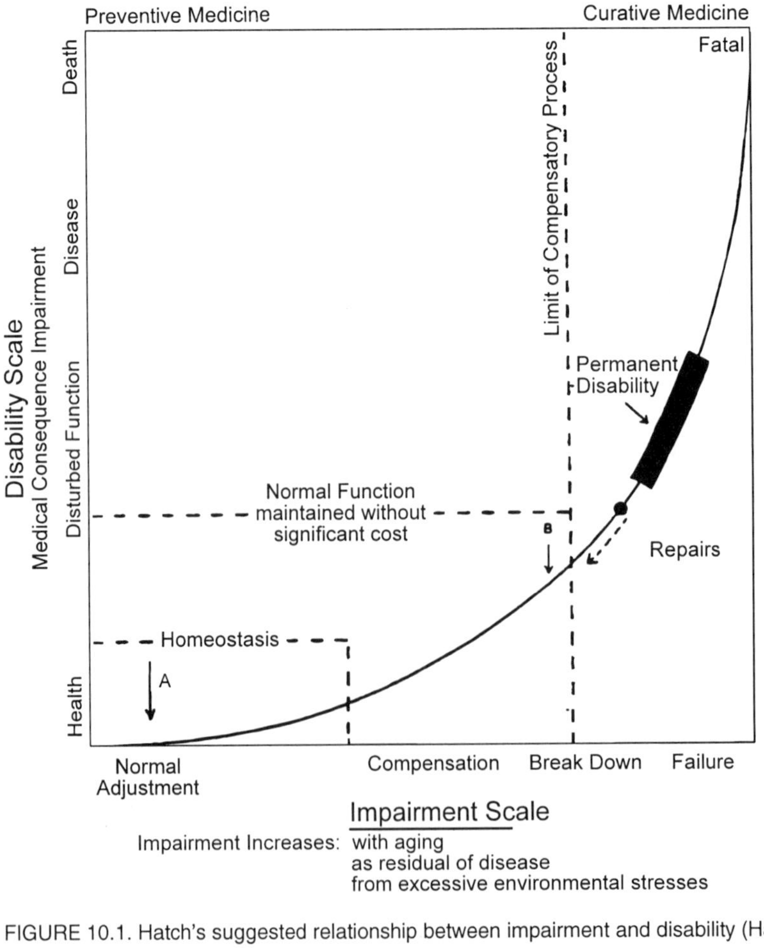

FIGURE 10.1. Hatch's suggested relationship between impairment and disability (Hatch, 1977).

At even higher values of the parameter, repair mechanisms will be unable to return the subject to a normal state, and permanent disability will occur. Finally, extreme values of the parameter will result in death.

10.5 DOSE–RESPONSE RELATIONSHIPS

The concept of a dose–response relationship implies that the level of response is a continuous function of dose or exposure. The available data are then used to delineate the detailed parameters of the function. Unfortunately, even when the general shape of the dose–response curve is chosen *a priori*, the available data frequently do not permit an accurate estimation of the quantitative parameters of the function such as the slope (in appropriate coordinates) or median. In both animal experiments and human epidemiological studies, only a few dose points can be identified; for example, some exposures with clearly adverse effects and perhaps only one with no effects (the *"no observed adverse effect level"* or NOAEL). Sometimes adverse effects are observed at the lowest measured exposure level (the "lowest observed adverse effect level" or LOAEL) but the NOAEL is unknown.

The term "no **observed** adverse effect level" highlights the fact that the observation of effects may be constrained by the size of the group of animals or humans under surveillance. It is likely that adverse effects will occur in a very small fraction of any population even at extremely low exposures but would be observable only in a very large group of exposed individuals. For instance, if a given exposure level actually causes an adverse effect in one person in a thousand, a risk which is usually considered to be quite significant, it may require a group of several thousand to observe this risk with a satisfactory level of statistical certainty. It is difficult and expensive to conduct epidemiological surveys on worker populations of this size and, for some substances, the total population with a potential for exposure may be too small to allow the calculation of a statistically significant risk level. Accordingly, it will probably never be possible to assure that no workers exposed at a given level will suffer an adverse effect. The best that can be achieved is that the risk will be reduced to some societally established "*acceptable risk.*" As reviewed by Vincent (1998), this raises issues which go far beyond science.

Where a substance is associated with several different adverse effects, it is frequently observed that they will have different dose–response relationships, NOAELs or LOAELs. The effect that occurs at the lowest exposure levels is identified as the "critical effect" and forms the basis for a TLV. The critical effect is not necessarily the effect of greatest health concern. Cancer and effects that adversely alter the reproductive process, shorten life or decrease the quality of life are properly of greater concern than transient or temporary nuisances. But if the latter occur at very low exposures, they may form the basis of the TLV. Formaldehyde is a good example of this distinction. The substance has been associated with cancers of the upper respiratory tract in animals and, in the U.S., OSHA has established a PEL of 1.0 ppm (since changed to 0.75 ppm). OSHA's quantitative risk assessment concluded that exposures below this level would carry incremental cancer risks of less than one in a thousand. However airborne concentrations of formaldehyde near 1.0 ppm are associated with a high prevalence of irritative symptoms of the eyes, nose and throat, particularly in unacclimatized individuals. Animal and human observations indicated that a ceiling TLV of 0.3 ppm would be necessary to bring the prevalence of irritative symptoms below 10% to 20% (ACGIH, 1992). While this proportion is higher than the goal of protecting "...*nearly all workers,*" the occurrence of persons hyperresponsive to formaldehyde make it difficult to identify a level that would protect the vast majority of workers. It was anticipated that this level would also reduce the potential risk of cancer.

Once a critical effect and a dose–response relationship have been identified, it is necessary to determine an exposure level at which adverse health effects will be prevented in "...*nearly all workers.*" ACGIH has not recommended a fraction of the population that corresponds to "...*nearly all workers,*" but quantitative risk evaluations of existing TLVs indicate that the average fraction for the majority of substances is near 0.001 (i.e., one in a thousand, although it should be noted that there are clear exceptions, as in the cases of lead, mercury, ozone, crystalline silica, etc.). Similarly, OSHA appears to use this risk level (OSHA, 1997). Particularly where cancer is the endpoint of concern, there are many quantitative risk assessment models for predicting risks for exposures below any that have been observed or tested. Some of these methods accommodate biological parameters such as lung clearance and aerosol dynamics in the lung, but many significant quantitative values are unknown or unmeasurable and so must be set on the basis of animal experiments and professional judgment. Different quantitative risk models can yield estimates of risk for the same exposure which differ by one or more orders of magnitude and it is frequently impossible to determine which model yields the "correct" risk estimate.

TABLE 10.4. Criteria for safety factors

Observations	Criteria
Health Effect	Serious health effects such as cancer or reproductive effects receive higher safety factors than transient nuisances.
NOAEL/LOAEL	A higher factor will be used where the lowest observed exposure is a LOAEL rather than a NOAEL.
Species	Higher factors are used where the critical data come from animal observations rather than humans in work situations.
Data Quality	Lower factors will be used when the critical data come from extensive high quality epidemiological studies.

In the absence of good quantitative estimates of risk or good exposure and outcome data, ACGIH will rely on safety factors based on the collective professional judgment of the Committee on Chemical Substances TLVs. These safety factors are applied to NOAELs and LOAELs, sometimes those observed with analogous compounds, but there is no set of "accepted" *a priori* safety factors. Safety factors for specific compounds will be based on a variety of criteria, as described in Table 10.4. A safety factor of only 1.0 may be used when a NOAEL is observed in humans for transient nuisance symptoms while factors in the range of one thousand to one million may be used where the best data are derived from small or poorly designed animal experiments. Some of the issues surrounding the use of such safety factors have recently been reviewed by Fairhurst (1995).

10.6 Establishing a TLV

The establishment of a TLV is far from a linear process. But two critical steps must be completed initially, namely the substance and the associated health effects in question must be clearly identified. During the evaluation of a working environment, air samples will be collected in the workplace and analyzed in a chemical laboratory or in the field with direct-reading instruments. The laboratory (or the direct-reading instrument) will analyze for the contaminant specified by the industrial hygienist. Accordingly, the substance analyzed must be the one that is associated with the disease which the hygienist is trying to control. Even where the element or the formula of the toxic substance has been correctly identified, it is important to recognize its detailed chemical nature. Chemical characteristics such as isomerism, oxidation state and crystal polymorphism must be taken into consideration.

Organic substances frequently occur in different isomeric forms, which are, in fact, different compounds with distinctly different toxic effects. Hexane (C_6H_{14}) has five isomers: normal hexane in which the carbon atoms in the molecule are in a single straight chain and four other isomers with varying branched chain configurations. Normal hexane is a severe neurotoxin causing permanent nerve damage, while the other isomers have not been associated with such effects. Different TLVs are assigned to the different isomers. In the case of inorganic compounds of metals, the metal may occur in different oxidation states (or valences), each with its own toxic effect. Chromium is a good example; it occurs with oxidation states of 2, 3 or 6 with the latter two being common in the work environment. Chromates [Cr(VI)] have been associated with lung cancer while trivalent chromium has not. Even where an element does not occur in different oxidation states, its compounds may occur in different crystalline forms with different toxicities. Silicon dioxide (SiO_2), which is probably the most common compound in the earth's crust, is a white or colorless (when pure) crystalline substance that occurs in several different crystalline and amorphous forms called isomorphs. The most common form is quartz, but other forms such as cristobalite and tridymite are also seen in the workplace. The latter two isomorphs behave as different substances with different physical and toxic properties.

Many substances are associated with several, sometimes disparate, adverse health effects. Each of these effects must be identified along with the part of the body affected. The respiratory tract deposition site and subsequent target organs will be used to identify the particle size fraction for which a TLV will be established, as described in previous chapters. Each effect will be associated with its own dose–response relationship and the same compound or element could be thought of as several different substances, each with its own target organ. For instance, cadmium causes carcinogenic effects in the lung and non-malignant effects in the liver and kidney, with different dose–response relationships for each effect and target organ. At the risk of making the system of OELs overly complicated, it might therefore be convenient to consider cadmium to be several different substances and then establish a separate TLV for each.

As discussed earlier, the majority of TLVs for substances occurring as aerosols are currently expressed in terms of 'total' aerosol. But it is anticipated that most such TLVs will eventually be expressed as inhalable particulate (where some, e.g., sulfuric acid mist, might be candidates for more appropriate expression in terms of the thoracic fraction). As inhalable samplers become more widely used, epidemiological studies will increasingly report exposures as inhalable particulate. However, the vast majority of aerosol exposures underpinning current TLVs were originally measured as 'total' aerosol, and it will be necessary to incorporate those measurements into the dose–response data that are used in the development of a TLV. A proposed procedure for accomplishing this is discussed below.

The procedure will be used for new and revised TLVs that will have completely new documentations. However, with more than 600 substances in the TLV list, there are only about 20 new documentations published each year for new and revised values. At that rate, it would be many years before all the TLVs for particulates could be reviewed and revised. In addition, for many substances there is no evidence that the current TLV is ineffective, and review of these substances will receive a low priority.

As noted earlier, the ratio of inhalable to 'total' particulate concentrations (using the IOM sampler and 37-mm cassette respectively, see Chapter 6) lies within the range from about 1 to 3, depending on the process which generated the dust. In Table 10.2, this range of ratios is divided into four recommended default ratios. In the future, the Chemical Substances TLV Committee will convert published exposures measured as 'total' aerosol to inhalable particulate using appropriate factors chosen from Table 10.2 for the dust-generating processes being reported. Then a dose–response relationship will be identified using doses or exposures all stated explicitly in terms of IPM. The outcome of this analysis will be an IPM-TLV.

Many substances have been assigned a TLV of 10 mg/m^3, the value associated with materials that were formerly called "*nuisance dusts*" for which no adverse health effects were observed except, perhaps, at extremely high dust concentrations. The TLV was not set to protect specifically against health effects, but rather to avoid visibility problems, enhance safety, and minimize employee discomfort. In some cases, the value was assigned as a default where animals exposed by non-airborne routes showed no adverse effects except at very high doses. In these cases, the TLV Committee will now use the default conversion factors in Table 10.2 to establish IPM-TLVs. Where available data do not permit the use of either of the above procedures, the 'total' aerosol TLV will remain in effect until more definitive observations have been published.

Most workplace environments contain several different hazardous substances and it will then be necessary to determine the concentration of each one in the air. If all of the components of the mixture have IPM-TLVs, it may be possible to take a single sample which is then analyzed for all the constituents, for instance by atomic absorption spectrophotometry. In the more general situation, it may be necessary to use several different samplers which collect different particle size fractions or which will be analyzed by different methods (e.g., gravimetric or chemical). Industrial hygienists have always been confronted with this situation, but the existence of four different particle size-selective TLVs ('total,' IPM, TPM and RPM) adds a further degree of complication. Where it had been possible to use a single 'total' aerosol sampler, it may now be necessary to use both an inhalable and a 'total' aerosol sampler to evaluate a single worker's environment. Similarly with a substance like cadmium which currently has both a 'total' aerosol and a RPM-TLV, two different samplers will be needed to evaluate exposures.

The ACGIH TLVs are very influential in the occupational standards-setting processes in many other countries. However, it is important to note that they may be described as "health-based" OELs because issues of technical feasibility and economics are not considered. By contrast, in the regulatory standards of the U.S., the U.K., Norway, Australia and elsewhere (Vincent, 1998), OELs are established by processes that are in large measure political. Issues of technical feasibility, economic cost, and competitiveness are often explicitly considered so that the official limits may be higher than health effects alone would suggest. On the other hand, there can be great differences between the OELs promulgated in various countries based on different perceptions of what are or are not significant health effects, what fraction of the population corresponds to "*...nearly every worker*," and the use of conservative assumptions in quantitative risk assessments. The differential treatment of carcinogens and non-carcinogens illuminates the perception of cancer as a dread disease which must be avoided at all costs while disabling or fatal diseases such as pulmonary fibrosis or kidney failure are treated far less stringently.

REFERENCES

American Conference of Governmental Industrial Hygienists (ACGIH) (1946), Proceedings of the Eighth Annual Meeting of the American Conference of Governmental Hygienists, April 7-13, Chicago, IL, pp. 54-55.

American Conference of Governmental Industrial Hygienists (ACGIH) (1968), *Threshold Limit Values for Airborne Contaminants,* ACGIH, Cincinnati, OH.

American Conference of Governmental Industrial Hygienists (ACGIH) (1992), *Documentation of the Threshold Limit Value for Formaldehyde*, ACGIH, Cincinnati, OH.

American Conference of Governmental Industrial Hygienists (ACGIH) (1993a), *Threshold Limit Values and Biological Exposure Indices*, ACGIH, Cincinnati, OH.

American Conference of Governmental Industrial Hygienists (ACGIH) (1993b), *Documentation of Threshold Limit Values and Biological Exposure Indices,* ACGIH, Cincinnati, OH.

American Conference of Governmental Industrial Hygienists (ACGIH) (1997), *Threshold Limit Values for Chemical Substances and Physical Agents: "Total", Inhalable, Thoracic and Respirable Matter*, ACGIH, Cincinnati, OH.

American Conference of Governmental Industrial Hygienists (ACGIH) (1999), *Threshold Limit Values and Biological Exposure Indices,* ACGIH, Cincinnati, OH.

American Industrial Hygiene Association (AIHA) (1996), *Emergency Response Planning Guidelines and Workplace Environmental Exposure Level Guides Handbook*, American Industrial Hygiene Association. Fairfax, VA.

Bartley, D.L., Chen, C., Song, R. and Fischbach, T.J. (1994), *Am. Ind. Hyg. Assoc. J.*, 55, pp. 1036-1046.

Bowditch, M. (1944), In setting threshold limits, Transactions of the Seventh Annual Meeting of the National Conference of Governmental Industrial Hygienists, May 9, 1944, St. Louis, MO, pp. 29-32.

Bowditch, M., Drinker, C.K., Drinker, P. et al. (1940), Code for safe concentrations of certain common toxic substances used in industry, *J. Ind. Hyg. Tox.* 22, p. 251.

British Medical Research Council (BMRC) (1952), Recommendations of the BMRC panels relating to selective sampling. From the minutes of a joint meeting of Panels 1, 2, and 3 held on March 4, 1952.

Comité Européen de Normalisation (CEN) (1992), Workplace atmospheres: size fraction definitions for measurement of airborne particles in the workplace, CEN Standard EN 481.

Commission of European Communities (CEC) (1978, 1982, 1990, 1991 and 1996), Directives 78/610/EEC, 82/605/EEC, 90/394/EEC, 91/322/EEC and 96/322/EEC, Directorate General Five (DGV), Luxembourg.

Cook, W.A., (1945), Maximum allowable concentrations of industrial atmospheric contaminants, *Ind. Med.*, 14, pp. 936-946.

Donaldson, K., Li, X.Y, and MacNee, W. (1999), Oxidative stress as a mechanism in ultrafine (nanometre) particle-mediated lung injury, *J. Aerosol Sci.*, 28, pp. 553-560.

Fairhurst, S. (1995), The uncertainty factor in the setting of occupational exposure standards, *Ann. Occup. Hyg.*, 39, pp. 375-385.

Hatch, T.F. (1977), Changing objectives in occupational health, *Am. Ind. Hyg. Ass. J.*, 38, pp. 425-431.

International Standards Organisation (ISO) (1992), Air quality — particle size fraction definitions for health-related sampling. ISO/TR 7708-1983 (E).

International Labour Office (ILO) (1991), *Occupational Exposure Limits for Airborne Toxic Substances*, International Labour Office, Geneva.

Izmerov, N.F., Sanotsky, I.V. and Sidorov, K.K. (1982), *Toximetric Parameters of Industrial Toxic Chemicals Under Single Exposure*, Meditsina Publishers, Moscow (1977), translation published by the United Nations Environment Program Centre of International Projects, Moscow (1982).

Kobert, R. (1912), Kompendium der praktischen toxicologie zum gebrauche fur arzte, *Studierende um Medizinalbeamte*, p. 45. Stuttgart.

LaNier, M.E. (Ed.), Threshold Limit Values — discussion and thirty-five year index with recommendations, *Ann. Am. Conf. of Govt. Ind. Hygienists*, 9, Cincinnati, OH.

MAK Commission (Senatskommission der Deutschen Forschungsgemeinschaft zur Prufung Gesundheitsschadlicher Arbeitsstoffe) (1997), Maximum Arbeitsplatz Koncentrationen, MAK Commission, Oberschleissheim, Germany, published by VCH (Verlag Chemie).

National Conference of Governmental Industrial Hygienists (NCGIH) (1942), Transactions of Fifth Annual Meeting of the National Conference of Governmental Industrial Hygienists, pp. 163-170.

National Institute for Occupational Safety and Health (NIOSH) (1997), *Pocket Guide to Chemical Hazards*, NIOSH, Cincinnati, OH.

Oberdorster, G. (1992), Respiratory tract dosimetry of particles: implication for setting of exposure concentrations and extrapolation modeling. In: Proceedings of WHO Conference in Hanover, Germany.

Orenstein, A.J. (Ed.) (1960), Recommendations adopted by the Pneumoconiosis Conference, In: *Proceedings of the Johannesburgh Conference* (A.J. Orenstein, Ed.), Churchill, London, pp. 619-621.

Santa Clara Center for Occupational Safety and Health (1995), *Health-Based Exposure Limits*, Santa Clara Center for Occupational Safety and Health, Santa Clara, California.

Sayers, R.R. and Dallavalle, J.M. (1935), Prevention of occupational diseases other than those that are caused by toxic dust. *Mechanical Engineer*, April 1935, pp. 13-17.

Terry, K.W. and Hewson, G.S. (1996), Comparative assessment of dust sampling heads used in Western Australian mines — implications for dust sampling practice, In: *Occupational Hygiene Solutions* (G.S. Hewson, Ed.), Proceedings of the 15th Annual Conference of the Australian Institute of Occupational Hygienists (Perth, December 1996), Australian Institute of Occupational Hygienists, Tullamarine, Victoria, Australia.

U.S. Department of Health and Human Services (1994), *The International Classification of Diseases, 9th Revision, Clinical Modification, Fifth Edition*, DHHS Publication No. (PHS) 94-1260. U.S. Department of Health and Human Services, Public Health Service, Health Care Financing Administration.

U.S. Occupational Safety and Health Administration (OSHA) (1997), Occupational exposure to methylene chloride, 62FR, pp. 1494-1619, January 10,1997.

Vincent, J.H. (1998), International exposure standards: a review and commentary, *Am. Ind. Hyg. Ass. J.*, 59:729-742

Walton, W.H. and Vincent, J.H. (1998), Aerosol measurement instrumentation in occupational hygiene: an historical review, *Aerosol Sci. Tech.*, 28, pp. 417-438.

Werner, M.A., Spear, T.M. and Vincent, J.H. (1996), Investigation into the impact of introducing workplace aerosol standards based on the inhalable fraction, *The Analyst*, 121, pp. 1207-1214.

PART II

EMERGING ISSUES IN PARTICLE SIZE-SELECTIVE AEROSOL SAMPLING

◆ ◆ ◆ ◆ ◆

Chapter 11

PARTICLE SIZE-SELECTIVE CRITERIA FOR DEPOSITED SUBMICROMETER PARTICLES

Michael A. McCawley

Centers for Disease Control, National Institute for Occupational Safety and Health, Morgantown, WV

11.1 Introduction

Current criteria for particle size-selective sampling, including those elaborated elsewhere in this book, are defined in terms of curves describing the probability that a particle of given aerodynamic diameter will penetrate into the particular region of interest of the respiratory system. Here it is important to note that "*penetration efficiency*" is defined as the ability of a particle to reach — but not necessarily deposit — in the region in question. In the conventions for inhalable particulate matter (IPM), thoracic particulate matter (TPM), and respirable particulate matter (RPM) discussed in previous chapters, this penetration efficiency for a particle with d_{ae} less than 1 μm is considered to be approximately 100%. Sampling of the mass concentration of any submicrometer particles with regard to these conventions therefore requires an efficiency as close to 100% as is practical.

This is not an unreasonable approach for silica, coal, and other mineral dusts like those encountered in the extraction industries, the primary particles for which the first particle size-selective criteria (for the respirable fraction) were originally established. Such dusts, associated with pneumoconiosis and other diseases of the deep lung, have widely dispersed particle size distributions with a large fraction of the respirable mass in the range of d_{ae} above 1 μm (Burkhart *et al.*, 1987; Seixas *et al.*, 1995). For those types of dust, it has been shown that a correlation exists between estimates of penetration and actual deposition (McCawley, 1993) and that, over the same range of particle size distributions, the mass contained in the TPM fraction correlates well with total lung deposition for particulate where the mass is predominantly in the supermicrometer size range.

When compared to the results obtained from 'total' aerosol sampling, it has become apparent that particle size-selective sampling along the lines advocated in this publication enables particulate control technology to be more effectively directed at those particle sizes that are likely to contribute to unhealthy exposures, either by eliminating the smaller particles or by allowing the production of predominantly larger ones. The use of 'penetration-based' criteria also enables the selection of relatively simple sampler pre-selectors (e.g., horizontal elutriators, cyclones, etc.). Such instruments, while not able to follow the lung deposition curves for submicrometer particulate, are capable of reasonable dose estimation for aerosol where the greater proportion of the mass resides in particles with d_{ae} greater than about 1 μm. This in turn has resulted in occupational exposure limits (OELs) that are better for health protection and more efficient in helping to direct the expenditure of resources for exposure control.

Notwithstanding the preceding, mineral dusts such as silica and coal are not the only dusts now recognized as candidates for size-selective exposure limits. As the list of aerosol-producing materials under consideration is expanded, so too is the need for consideration of other criteria for particle size-selective sampling. Not all sources produce particles predominantly greater than 1 μm. It has long been recognized that the 'coarse' or 'mechanical' fraction of aerosol is produced, as the latter connotation for the fraction implies, largely by mechanical processes. The energy limitations on mechanical processes such as cutting, grinding and crushing tend to produce particles greater than 1 μm.

Processes such as evaporation and condensation, on the other hand, are not so limited and so can produce aerosols with a very different size range. Many such processes tend to generate very small primary particles which, though they may grow by aggregation, usually remain smaller than 1 μm. This is well known for processes such as combustion, welding, spraying and even nuclear decay. $PM_{2.5}$, sometimes termed "*fine*" particulate matter (see Chapter 5), represents the finest particulate size-selective sampling criteria previously published. There are definitions for "*ultrafine*" particulate in the literature, specifying it to refer to particles with geometrical diameter less than either 0.2 or 0.1 μm; however, this definition is not based on any rationale relating to lung physiology, aerosol penetration or deposition (see Chapter 12).

A primary goal of previously published size-selective sampling criteria was to provide a means of assessing human respiratory tract particulate exposure, with the underlying supposition that this would relate, in some way, to dose.* But, strictly speaking, deposition is a necessary step in establishing a physical link between the exposure measurement and the particulate available to the lung surface as dose.

This might not necessarily be an issue when sampling for compliance purposes; however, it becomes more relevant in other applications of health-related aerosol sampling. In epidemiology, for example, where it is the intention to directly link exposure with a disease outcome, it is implied that sampling is being carried out in order to assess what is **deposited**. While there has been some debate about how well penetration serves as a surrogate for mass deposition for relatively coarse aerosol (Soderholm and McCawley, 1991; Hewett, 1991), it is clear that this is certainly not the case for relatively fine aerosols where the highest proportion of the mass is in the submicrometer range.

In this chapter, an argument is presented that articulates the need, in some exposure assessment situations, for sampling to reflect what is actually deposited in the lung, in contrast to what penetrates to the lung. This leads to an extension of sampling criteria beyond those based on the concept of particle penetration.

11.2 Penetration versus deposition

The essence of past arguments is that penetration of a particle to a lung region is a necessary step in the exposure process. Measurement of penetration according to a particle size-selective criteria is assumed to be more closely related to dose and therefore more useful in protecting health than simply measuring the 'total' amount of particles in the air outside the body. However, that argument is flawed if there is not a direct relationship between penetration and deposition because deposition is a necessary step in establishing dose. Although, as already mentioned, that direct relationship has been established for widely dispersed dusts, penetration for more finely dispersed

*Here the term "dose" is used to refer to what is actually deposited.

particles is close to 100% for all particle sizes below about d_{ae} = 1 μm* while deposition continues to vary sharply. It may be estimated that variability in the particle size distribution of submicrometer aerosols could lead to a range of differences of as much as 400% between current penetration-based measurements and the estimate of actual mass deposition (McCawley, 1993). While it might appear that criteria based on a penetration curve are more protective of a person's health and thus offer a more conservative estimate of exposure, this is not true in every case for submicrometer aerosols. By way of illustration, consider the following three hypothetical situations:

1. Respirable dust sampling is used to estimate diesel exhaust particulate exposure in an underground mine. Diesel equipment in use during the sampling produces a particle size distribution with a median particle diameter ($_{50}d_{ae}$) of 0.4 μm and a geometric standard deviation of 1.5. The measured respirable dust concentration is 0.50 mg/m^3. But the concentration which relates to the actual dose delivered to the lungs is approximately 0.10 mg/m^3. New equipment is purchased that then reduces the measured respirable dust concentration to 0.25 mg/m^3. However, the new equipment produces exhaust particulate with $_{50}d_{ae}$ = 0.08 μm and a geometric standard deviation of 1.5. The delivered dose to the lungs now corresponds to 0.20 mg/m^3, double what it had been with the old equipment. So, in this example, while there is a two-fold decrease in the concentration of respirable dust, this is more than offset by a four-fold **increase** in the deposition efficiency, resulting in the undetected doubling of delivered dose for the exposed workers.
2. $PM_{2.5}$ monitors are used to measure ambient air concentrations near lead smelting operations in two different towns. In both towns, the airborne lead concentration is measured to be the same at 4 μg/m^3. But the particle size distributions are distinctly different. One has a $_{50}d_{ae}$ value of 0.3 μm and the other a $_{50}d_{ae}$ value of 0.03 μm, while both have geometric standard deviations of 2. People exposed to the smaller particle size could be expected to have a 200% higher dose potentially accumulating as much as 6 μg of lead per deciliter of blood per week (Bridibord,

*Here it is acknowledged that the formal definition of particle aerodynamic diameter, based as it is on the falling speed of a particle under gravity and how that compares to that for a sphere of density 10^3 kg/m^3, may be less appropriate for particles sizes below 1 μm. In this chapter, therefore, the nomenclature "d_{ae}" is used to refer to a particle generally equivalent to a sphere of density 10^3 kg/m^3.

1977) more than in the other community with the same measured $PM_{2.5}$ ambient concentration.

3. A more complicated situation could occur if epidemiologists use a cohort of miners and records of respirable dust levels in a metal ore mine to set an RPM-based OEL for a metal-containing dust. Assume the $_{50}d_{ae}$ value for particles in the mine is 2.0 μm, with a geometric standard deviation of 3. This OEL is then used for assessing exposures in the subsequent refining operation for the metal where now the aerosol appears as a fume with a $_{50}d_{ae}$ value of 0.02 μm and a geometric standard deviation of 3. For respirable dust sampling for coarse size range dusts, there is a relatively unchanging ratio between the penetration and deposition of approximately 3 to 4. Thus the actual deposited dose of dust is constantly three to four times less than the measured amount only so long as the $_{50}d_{ae}$ value for the particles is larger than about 1 μm. As long as workers are exposed to the supermicrometer aerosol, regardless of how the size distribution changes, it is fair to assume that they will receive the same protection from the respirable dust standard. However, if the same RPM-based OEL is used in relation to metal fumes in a refining operation, where the $_{50}d_{ae}$ value for the aerosol will certainly be below 1 μm, the factor relating the RPM-based exposure measurement to the actual deposited dose will change to only 1.5. In this case, the workers exposed to the fume will receive twice the deposited dose of the workers exposed to the dust, even though the measured exposure is the same. Thus, they will not be equally protected by the standard. This points out the anomaly that can occur with such RPM-based standards. In this example, the epidemiologists had used the measurement of particles which were mainly in the supermicrometer size range to generate an OEL which, if applied for the same substance in a new situation, might not achieve the same level of actual protection of workers.

In the first two examples, it is demonstrated that, when there is not a constant proportion between penetration and deposition, penetration can not be assumed to be inherently more conservative. The likelihood that the particle size distribution can change is also not small. The simplest example of this would be the variability in particle size from one operation to the next: in the case of diesels, from one type of engine to another; in the case of welding fume, from one type of welding process to another; in the case of ambient air, from one location to another; and so on.

In the third example, an additional concern is raised that a penetration-based size-selective sampling exposure standard may not be equally protective in all cases. If the information used to set an OEL for fine particulate matter does not adequately account for the range of size distributions that could be encountered, the OEL could be set too low. For this reason and those stated above, a health-based, particle size-selective criterion for submicrometer particulate that is not directly proportional to deposition is not desirable.

It has been suggested that sampling for diesel particulate in underground coal mines in the presence of coal mine dust should be carried out using an impactor with a cut point at $d_{ae} = 1$ μm (McCawley and Cocalis, 1986). In this specific case, the rationale was that separation of the diesel particulate from coal mine dust could be achieved on a particle size-selective basis because the coal mine dust particle size distribution had been well characterized as mechanical-mode size dust with 90% of the mass greater than 1 μm (Burkhart *et al.*, 1987; Seixas *et al.*, 1995). This approach assumed that diesel particulate would then account for any particulate mass found in the fraction of dust below 1 μm. The intent of this proposal was very different from that for established particle size-selective criteria. The particle size selection was performed in order to differentiate between two distinct classes of particles, for which an established chemical separation technique did not exist at the time. This approach is in marked contrast with the rationale underlying the particle size-selective sampling criteria described elsewhere in this publication.

The question is: are the above concerns about fine particles justified in terms of the number of cases where the conventional penetration-based criteria might not be applicable? The answer is "yes." A recent review (Fisher and McCawley, 1997) of published ambient air size distributions indicates a major percentage of those size distributions include submicrometer-size particulate to a measurable extent (see Figure 11.1). In occupational environments, both diesel exhaust and welding fumes have been extensively characterized and found to be predominantly submicrometer in size (Cantrell and Rubow, 1992; Hewett, 1995). An association between mortality in the general population and fine particulate mass concentration (as defined in terms of $PM_{2.5}$) has also been noted, with the added observation that much of the $PM_{2.5}$ mass may be due largely to the mass of the submicrometer fraction (Stahlhofen *et al.*, 1989).

In trying to extend the concept of particle size-selective sampling to cover smaller particle sizes, two distinct concerns must be addressed. The first is that consideration should be given to an alterna-

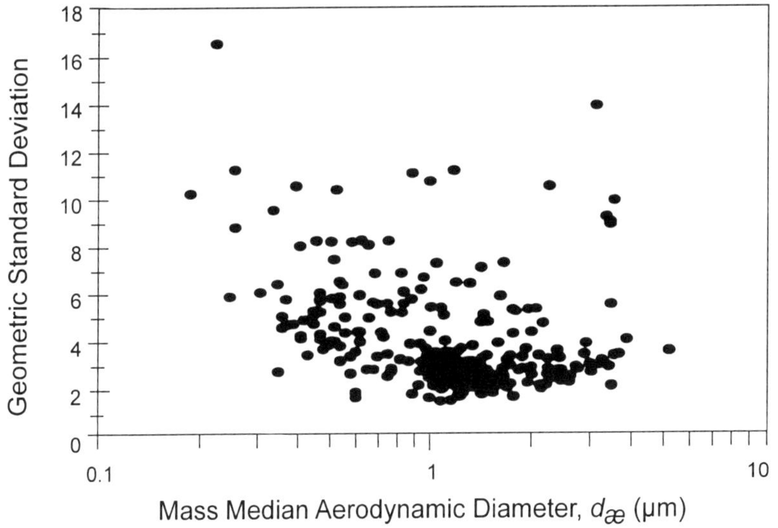

FIGURE 11.1. A compilation of published ambient air particle size distributions (by mass) showing the high frequency of submicrometer median diameters (Fisher and McCawley, 1997).

tive approach to particle penetration in the lung as the basis for the size-selective sampling criteria for smaller particles. The second is that due consideration should be given in each practical situation to adopting the most appropriate metric of dose and to the possibility that the basis of that dose may not always necessarily be particulate mass, especially when considering very fine particles.

11.3 A DEPOSITION-BASED CRITERION FOR SUBMICROMETER PARTICLE SAMPLING

In an extension of the traditional penetration-based rationale for particle size-selective sampling criteria, a further criterion may be appropriate for some situations where it is defined in terms of lung deposition. The new criterion, which may be termed "*deposited submicrometer particulate*" (DSP), follows the pattern for total deposition of particles in the lung and therefore is expressed in terms of a spherical particle with a density of 10^3 kg/m^3 (see Figure 11.2). For this, an oral breathing frequency of 15 breaths per minute and a tidal volume of 1.45 L were chosen to represent a normal work rate. The DSP curve also provides a reasonable match with the lung deposition curve for a resting nasal breathing rate of 7 breaths per minute with a tidal capacity of 0.5 L.

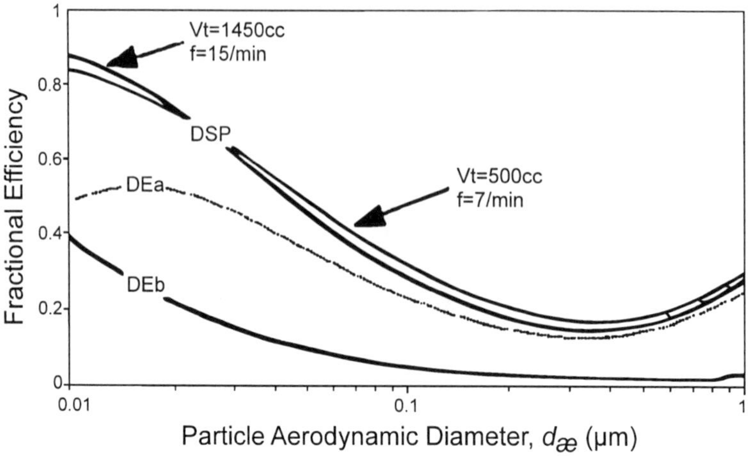

FIGURE 11.2. The criterion for *deposited submicrometer particulate* (DSP). Also shown are the corresponding curves for tracheobronchial (DE_b) and alveolar (DE_a) deposition for oral breathing with a tidal volume of 1.45 L at a frequency of 15 breaths/min (corresponding to 'at work') and a tidal volume of 0.5 L at a frequency of 7 breaths/min (corresponding to 'at rest'), as functions of d_{ae} as calculated for spheres with density 10^3 kg/m^3 (Stahlhofen *et al.*, 1989).

The DSP curve may also be used as a basis for OELs for submicrometer aerosol since there is a relatively constant proportion between DSP and the deposition of aerosol in the thoracic or total lung region (DE_{ba}). In fact, there is no correction required because the two are approximately equal. There is also a constant proportion for all but the smallest particle sizes in the gas exchange region (DE_a) (where DE_a is equal to 0.75 DSP) (see Figure 11.3). The DSP fraction can be expressed as an empirical function of d_{ae}, thus

$$\mathrm{DSP} = 1 - \{1.03 \exp\left[-(\log_{10} d_{ae} + 0.49)^2/1.77\right] - 0.18\} \qquad \mathbf{11.1}$$

for d_{ae} up to 1 µm, where d_{ae} for particles in this size range is the diameter of a spherical particle with density 10^3 kg/m^3 (as distinct from aerodynamic diameter, d_{ae}, as defined strictly in terms of settling velocity). DSP is not defined for particles with d_{ae} larger than 1 µm.

11.4 Sampling considerations

The obvious choice for sampling particulate according to the DSP criterion given above would be to use a sampler that pre-separated the large particles aerodynamically with sharply increasing penetration efficiency below d_{ae} = 1 µm and then allowed either penetration and

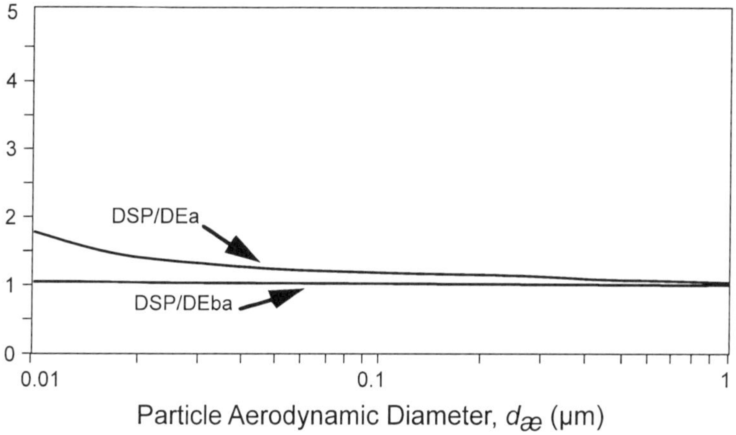

FIGURE 11.3. Comparison of the DSP to thoracic deposition (DE_{ba}) or alveolar deposition (DE_a), expressed as DSP/DE_{ba} and DSP/DE_a respectively.

detection or direct collection (the latter being the more likely) of the particles based on their size. The shallow slope of the DSP curve could be matched with a single-stage multi-screen diffusion cell (see Figure 11.4). As an example, this can be accomplished with an instrument design that is currently available. Figure 11.4 refers to a cell with 10 screens in series, as is found in the TSI Model 3040 diffusion battery described by Cheng *et al.* (1980). If higher flowrates are desired, the additional number of screens required to meet the operating charac-

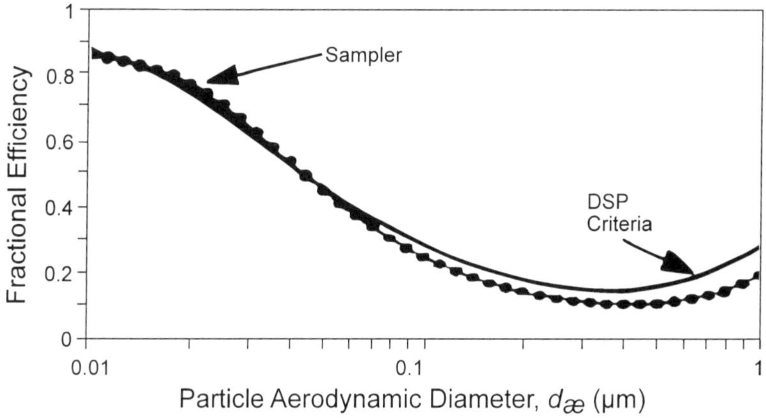

FIGURE 11.4. Collection efficiency of a single-stage, multi-screen diffusion cell (Stahlhofen *et al.*, 1989) compared to the DSP curve.

teristics may produce a pressure drop that would possibly preclude the use of personal sampling pumps. For area (or static) sampling in ambient air studies, however, this may not be a problem. Nor may it present a problem in determining the effect of changes in control technology, where area samplers offer a clearer picture of changes in the general environment.

An alternative approach to sampling may be to characterize the particle size distribution of the environment and correct personal exposure measurements (e.g., based on TPM or RPM) based on that information. For that matter, for any size-selective sampling criterion to be properly applied, the environment should be well characterized. This characterization includes knowledge of the particle size distribution. With a knowledge of the size distribution of the DSP particulate, the fractional collection efficiency factor can be calculated and applied to obtain a measure of the total particulate — whether by mass, surface area or count — which could be collected as a personal exposure measurement. Figure 11.5 allows this to be done and applied to the DSP, using the geometric mean and geometric standard deviation of the measured particle size distribution. Thus there are, in fact, two alternatives: firstly, to use an instrument with particle size-selection characteristics matching those of the DSP curve; secondly, to determine the particle size distribution and apply the appropriate collection efficiency factor. In addition, it should be noted that knowledge of the particle size distribution should be required for any

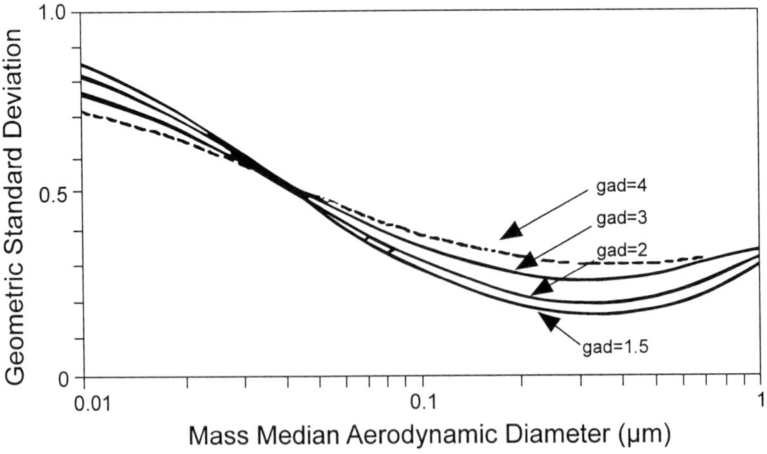

FIGURE 11.5. Correction factor for 'total' aerosol measurements to obtain the fractional deposition amounts for 'at work' oral breathing (see Figure 11.2).

cited exposure assessment studies used to set an OEL based on the DSP criterion.

One final issue that needs to be mentioned is whether the actual mechanism of injury and subsequent disease is best linked to dose as it relates to either particle mass, surface area or number. This is bound to affect the analytical method chosen to quantitate collected samples. The issue could be readily dismissed by saying that any decision on the analytical method, in relation to DSP or to any other of the criteria described in this book, does not really affect the specification of the size-selection criteria. However, it is worthy of discussion because selection of the sampling method might well be determined by the analytical method. Most OELs have, until now, been expressed in terms of mass. But had it been known that particle surface area or number was a better metric, that information would have affected the OEL itself, but not its underlying the particle size-selection criterion. As mentioned above, it may be indeed possible to derive an exposure–response relationship for fine particulate using a mass measurement for exposure. However, recent work by Peters *et al.* (1997) suggests that the use of particle number concentration yields a better correlation between exposure and disease for populations exposed to ambient air than does fine particulate mass. This may be because particle number may be more indicative of the effective toxic dose of fine particles. This same report goes on to theorize on the toxicologic mechanisms; cellular reactions in both the bronchial and alveolar regions of the lung are cited as possible factors, as well as the fact that exposure to ultrafine particulate may be associated with impairment of the clearance mechanism of the lung. For the latter, particles retained for longer periods may be more likely to move through the pulmonary epithelium thus making them available for retention in the interstitium (Ferin *et al.*, 1990; Oberdorster *et al.*, 1995). But, as noted above, there is no known biological significance to a delineation point of 0.2 μm or 0.1 μm for defining ultrafine aerosol. Recent epidemiologic and biological studies that associate ultrafine particles with health outcomes were not designed to develop exact specifications for particle size-selection criteria and have generally used the term “ultrafine” as a convenient way of saying that the particle sizes of interest were some subfraction of the fine particulate below about the 2.5 μm embodied in the $PM_{2.5}$ criterion. Studies aimed at setting exposure limits for DSP need to be aware of this. Until those studies are done, there can be no general recommendation for the most appropriate exposure metric, whether it be mass, surface area or number concentration. It is entirely possible that different substances may have different metrics and therefore all three may find some value.

11.5 Conclusion

The DSP criterion is based on an approach which is different from the penetration-based rationales of IPM, TPM, and RPM. It aims to provide an exposure measurement guideline based on total lung deposition for particles in the range of particle diameter (d_{ae}) from 0.01 μm to 1.0 μm. This is a useful extension of the penetration-based criteria because it provides a framework — that might be applicable for some aerosol exposure situations — for measuring exposure in a manner that reflects what is actually deposited in the lung independently of the particle size distribution. For example, it can be shown that using any of the previously published criteria for RPM may result in differences of as much as 400% between exposure and the potential dose to the lung surface.

A normal range of breathing patterns encompassing rest and moderate work levels has a negligible effect on the deposition curve, so a single DSP curve is defined. It can be matched, in principle, with a multiple-screen pre-selector as a sampling device which employs a combination of impaction, sedimentation, and diffusion (as would the lung itself), to collect particles over the size range of interest. This type of device could be modified to allow analysis of the sample for number, surface area or mass of the aerosol being particle size-selectively sampled.

It is assumed that the DSP curve might be used as the basis of OELs for fumes, combustion aerosol and other assorted fine particulate. As such it has the potential for providing a more accurate measure of exposure for aerosols like those mentioned, and therefore a scientifically sounder basis for control. However, it is a concept which is at present in its infancy, and it needs to be further explored and validated before it can be deployed in actual OELs.

References

Bridbord, K. (1977), *Human exposure to lead from motor vehicle emissions*, DHEW (NIOSH) Publication Number 77-145.

Burkhart, J.E., McCawley, M.A. and Wheeler, R.W. (1987), Particle size distributions in underground coal mines, *Am. Ind. Hyg. Assoc. J.*, 48, pp. 122-126.

Cantrell, B.K. and Rubow, K.L. (1992), Diesel exhaust aerosol measurements in underground metal and nonmetal mines, In: *Diesels in Underground Mines:*

Measurement and Control of Particulate Emissions, Bureau of Mines IC 9324, Washington DC.

Cheng, Y.S., Keating, J.A. and Kanapilly, G.M. (1980), Theory and calibration of a screen-type diffusion battery, *J. Aerosol Sci.*, 11, pp. 549–556.

Ferin, J., Oberdorster, G., Penney, D.P. *et al.* (1990), Increased pulmonary toxicity of ultrafine particles? 1. Particle clearance, translocation, morphology, *J. Aerosol Sci.*, 21, pp. 381-384.

Fisher D.L. and McCawley, M.A. (1997), Lack of correlation between PM10 measurements and upper respiratory tract dose, In: *Inhaled Particles VIII* (N. Cherry and T.L. Ogden, Eds.), Pergamon Press, Oxford, U.K., pp. 14-18.

Hewett, P. (1991), Limitations in the use of particle size-selective sampling criteria in occupational epidemiology, *Appl. Occup. Environ. Hyg.*, 6, pp. 290-300.

Hewett, P. (1995), The particle size distribution, density and specific surface aea of welding fumes from SMAW and GMAW mild and stainless steel consumables, *Am. Ind. Hyg. Assoc. J.*, 56, pp. 118-135.

McCawley, M.A. (1993), Caveats in the use of particle size-selective sampling criteria, In: *Proceedings of the Second International Conference on Occupational Health and Safety in the Minerals Industry,* Perth, Australia, March 1993.

McCawley, M.A.and Cocalis, J.C. (1986), Diesel particulate measurement techniques for use with ventilation control strategies in underground coal mines, *Ann. ACGIH,* 14, pp. 271–281.

Oberdorster, G., Gelein, R.M.., Ferin, J. *et al.* (1995), Association of particulate air pollution and acute mortality: involvement of ultrafine particles? *Inhal. Toxicol.*, 7, pp. 111-124.

Peters, A.,Wichmann, A., Tuch, T. *et al.* (1997), Respiratory effects are associated with the number of ultrafine particles, *Am. J. Respir. Care Med.*, 155, 1376-1383.

Seixas, N.S., Hewett, P., Robins, T.G. *et al.* (1995), Variability of particle size-specific fractions of personal coal mine dust exposures, *Am. Ind. Hyg. Assoc. J.*, 56, pp. 243-250.

Soderholm, S.C. and McCawley, M.A. (1991), Should Dust Samplers Mimic Human Lung Deposition? *Appl. Occup. Environ. Hyg.*, 5, pp. 829-835.

Stahlhofen, W., Rudolf, G. and James, A.C. (1989), Intercomparison of experimental regional aerosol deposition data, *J. Aerosol Med.*, 2, pp. 285-308. (1989).

◆ ◆ ◆ ◆ ◆

Chapter 12

HEALTH-RELATED MEASUREMENT OF VERY, VERY SMALL PARTICLES

James H. Vincent

Department of Environmental Health Sciences, School of Public Health, University of Michigan, Ann Arbor

12.1 PHYSICAL BACKGROUND

There was a time when fine particles were considered to be those which, after inhalation, could penetrate down to the alveolar region of the lung — what is now referred to as "*respirable aerosol.*" More recently interest has been drawn towards still finer particle fractions, not least because of their possible association with health effects specific to those sizes of particle. The term "*ultrafine*" is used by many to refer to particles whose diameters are less than 0.1 μm, and the use of that term seems to suggest that such particles represent the ultimate in 'smallness.' However, interest is starting to focus on particles that are still finer yet; this is stimulated in part by some of the technology-oriented aerosol research which is beginning to yield insights into the physical nature of such particles and their reactivity (Pui *et al.*, 1998). Here, the central issue concerns the question about whether it is particle number count which is the most appropriate metric relating exposure to health effects, as opposed to the particle mass which underpins most considerations of larger particles.

It is noted that a 0.1 μm ultrafine particle measures 100 nm on the nanometer scale. By contrast, a typical atom or molecule has dimensions of about 0.3 to 0.5 nm. This depends on the atom or molecule in question and on its electronic state. The latter arises from the fact that, at the atomic or molecular level, the size of the entity is determined by the probability densities of the orbiting electrons. In turn, therefore, a molecule which is excited will have a larger characteristic size by virtue of the larger radius of those orbiting electrons. What is important in relation to particles is that as they become smaller and smaller, approaching the order of magnitude of the molecules themselves (say, a few nm), their physical state becomes distinctly different from that of larger particles from ultrafine upwards. So in order to refer to such particles as a size fraction in its own right, the term "*very, very small*" particles has been coined.

For a particle which is very, very small, the number of molecules or atoms the particle contains becomes so few that a high proportion of them lie at the surface of the particle. For example, a 20-nm particle has 12% of its molecules at the surface, and a 10-nm particle has 25% at its surface (Preining, 1998). At this level, therefore, even the concept of a "surface" has limited meaning because the material structure of the particle can no longer be regarded as a continuum. In turn, the material itself can no longer be thought of simply as "solid" or "liquid," and we enter the world of "cluster physics." Here, the magnitude and spatial distribution of the surface reactivity of a particle become strongly influenced by the way in which the quantum mechanical natures of the individual molecules combine in the overall particulate entity. Such considerations may become very important in relation to how such particles interact with other molecules and particles. This is what has stimulated the great current interest in the study of nanometer-sized aerosol particles in reactive systems and their applications in the synthesis of new materials. It also provides much food for thought in relation to how such particles might interact with biological organisms or cells, and hence on their potential toxicological effects.

12.2 HEALTH EFFECTS AND A POSSIBLE BASIS FOR A CRITERION

General interest in health effects of fine particles comes from the growing awareness of health effects associated with exposures to particles finer than those contained just within the $PM_{2.5}$ and PM_{10} fractions which underpin current Environmental Protection Agency (EPA) air quality regulations in the United States. This interest is being driven by current concerns that such particles in ambient air, as yet poorly defined and whose properties are as yet poorly understood, may be associated with the observed increases in mortality linked to cardiopulmonary disease in vulnerable populations (see also Chapter 5) (Dockery *et al.*, 1993; Pope *et al.*, 1995).

What has triggered this new interest? Evidence is emerging to support the view that very, very small particles are very important in relation to health, perhaps increasingly so. For example, Kittelson (1998) has noted that, although the latest generations of diesel and spark engines are improved in terms of the **mass** of particulate emissions, the **number** of nano-sized particles emitted has increased sharply. This may also be true for emissions from aircraft engines. Little evidence about the possible health effects of such small particles is available in relation to occupational exposures. But attention is caught by one striking example of a worker who died of *acute respiratory distress syndrome* (ARDS) following a brief exposure to metal fume during nickel spraying using a thermal arc process (Rendall *et al.*, 1993). From studies in a subsequent reconstruction of the exposure, the results showed that there could have been exposures to very high number concentrations of particles less than 50 nm in diameter. Measurements of nickel in urine taken before the worker died revealed a level which was "... *excessively high*." Since the species of nickel aerosol in question was not expected to be of a highly soluble form (probably metallic or oxidic), the possibility is raised that the very small size of the particles may have played a significant role in the worker's death.

Some basic biological research has been conducted to ascertain the specific characteristics of very fine particles, in the size range from about 10 nm (0.01 μm) up to about 500 nm (0.5 μm), in relation to what makes very fine particles more toxic to biological systems than larger particles of the same material. For example, Oberdorster *et al.* (1992 and 1995) and Donaldson *et al.* (1998) have carried out animal studies for particles of teflon and titanium dioxide respectively, and showed that particles of diameter less than about 50 nm (0.05 μm) produced much stronger inflammatory responses than 500-nm (0.5-μm) particles at the same mass concentrations.

Chen *et al.* (1995) provided important evidence regarding the role of acid particle number in cellular response. Guinea pigs were exposed to varying amounts of sulfuric acid layered onto 10^8 ultrafine carbon core particles and to a constant concentration of acid (300 $\mu g/m^3$) layered onto 10^6, 10^7, and 10^8 carbon core particles respectively. All of these particles had diameters of approximately 90 nm. Indicators of irritancy potency on macrophages harvested from the lungs of exposed animals clearly showed an increased response to a constant dose of acid when it was divided into an increased number of particles, as well as a response to increased dose of acid at a constant particle number concentration.

Seaton and colleagues (1995 and 1996) considered the possible causative factors linking exposure and the observed mortality in the general population. They proposed a hypothesis in which exposure to very, very small particles in the size range around 50 nm

> *"... characteristic of air pollution (may) provoke alveolar inflammation leading to acute changes in blood coagulability and release of mediators able to provoke attacks of acute respiratory illness in susceptible individuals. The blood changes result in an increase in the exposed population's susceptibility to acute episodes of cardiovascular disease; the most susceptible suffer the most. This hypothesis, being based on the number, composition and size — rather than on the mass — of particles accounts for the observed epidemiological relations."*

It is clear that there is a need for further research to confirm the plausibility of such a hypothesis or to provide a basis for new ones. In the meantime, on the basis of what is already known, we are not yet in a position to suggest a specific particle size-selective criterion by which such exposures might be measured. However, the available evidence points towards the possibility that an appropriate health-relevant metric might be expressed in terms of the number concentrations of very, very small particles in the range below a few — or at most a few 10s — of nm.

12.3 Measurement options for very, very small particles

Measurement of very, very small particles like those described requires instrumentation that departs significantly from the sampling approaches that have been traditionally applied in industrial hygiene and which underpin the particle size-selective criteria that are featured elsewhere in this book. Because particle number is expected to be the most appropriate metric for concentration, and because particle mass concentrations for aerosols with particles in the size range of interest are extremely low, it is expected that sampling particles onto filters, screens or substrates for subsequent assay may not be the most promising approach for routine measurement purposes. Instead, we can turn to direct-reading instruments of the type which have been developed in aerosol research laboratories in recent years as interest in nano-sized particles has surged for other reasons (as mentioned earlier). The ones most likely to be useful in the context described are those which can selectively respond to and count very, very small particles in the size range of interest.

DIFFERENTIAL MOBILITY ANALYZERS

The *differential mobility analyzer* (DMA) (Knutson and Whitby, 1975) responds to the electrical mobility of a particle from which, if the charge carried is known or can be inferred, the particle size can be determined. A typical configuration is shown in Figure 12.1 in which the sampled aerosol is charged unipolarly and then is admitted into the edge of an annular space inside a cylindrical cell, winnowed by a sheath of clean air (Vincent, 1995; Pui, 1996). A high voltage is applied between the outer and inner electrodes of this annular system, producing an electric field that causes the particles to migrate towards the central electrode. For a given voltage, particles of a defined electrical mobility pass through the exit slit and into a condensation particle counter (CPC, see below) by which the selected particles are grown to a size at which they can be detected and counted by optical scattering. In this system, for particles that are small enough, it can be ensured that each particle collected in this way carries only one electric charge — in which case it is a simple matter to determine the particle's physical size. By scanning through a range of voltages to the central electrode, a complete particle size distribution is obtained over the specific size range of interest. From such information, the number of sampled particles within any given particle size range can be determined.

Although conventional versions of the DMA (like those currently commercially available from TSI Inc., St. Paul, MN) work well for counting commercially particles in the range from 500 nm down to about 20 nm, resolution at the low end has been limited by "*diffusive broadening*." Here, the particles no longer follow exact, deterministic trajectories. However recently improved versions of the basic DMA concept by Chen *et al.* (1998) have yielded the "*Nano-DMA*" which can accurately select and count particles down to 3 nm. Such instruments are expected to be commercially available in the near future, driven not least by potential applications in the health-related measurement of very, very small particles.

CONDENSATION PARTICLE COUNTERS

The principle of the *condensation particle counter* (CPC) (Agarwal and Sem, 1980) involves first the sampling of fine particles, raising the

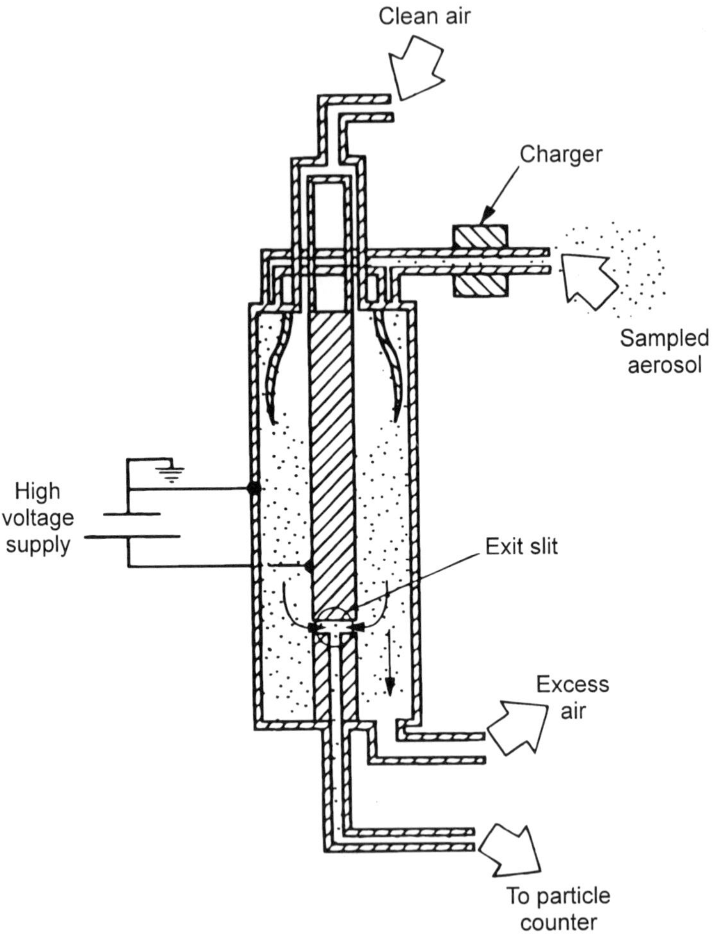

FIGURE 12.1. Diagram to illustrate the principle of the differential mobility analyzer (DMA)

sampled air to an elevated temperature and then introducing it into a region which is saturated, usually with alcohol. The atmosphere is then made to supersaturate by cooling in a condenser, leading to condensation of the vapor molecules onto the particles, causing the particles to grow to the point where they are large enough to be detected optically. The principle of operation is shown in Figure 12.2.

The particle size limit of earlier versions of this instrument at the low end arose because the smaller sampled particles were not able to grow during their time in the condenser to the point where they could be detected sufficiently accurately. With this in mind, Stolzenberg and

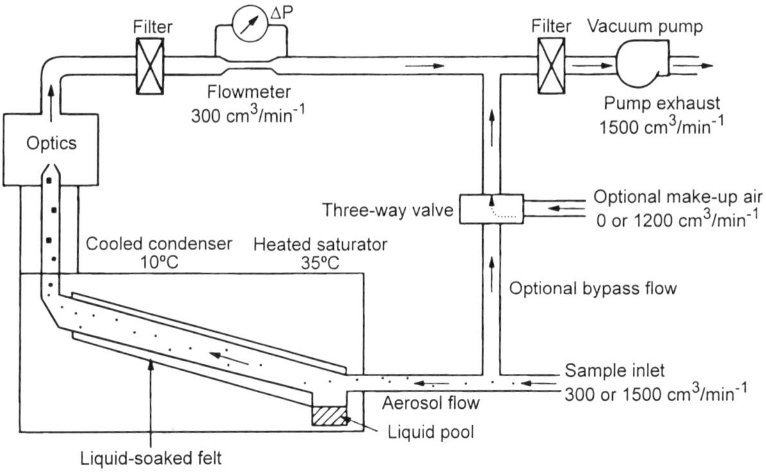

FIGURE 12.2. Diagram to illustrate the principle of the condensation particle counter (CPC).

McMurry (1995) developed an *ultrafine condensation particle counter* (UCPC) in which the aerosol flow in the condenser is constrained to the condenser centerline by means of a clean-air sheath flow so that particles of all sizes encounter the same variation of saturation ratio. By means of this instrument, all particles of diameter about 10 nm and above grow to about the same size so that, although they can be counted, they cannot be sized. However, in the range below 10 nm, particles grow to a size that is dependent on their initial size. Weber *et al.* (1998) have recently shown that optical counter pulse height information from the UCPC can be inverted to provide accurate particle size distributions in the range from 3 to 10 nm. Indeed, it is predicted that such an instrument may be capable of sizing and counting particles of size down to about 1 nm.

Versions of both the basic CPC and the UCPC are available from TSI Inc., St. Paul, MN.

OTHER APPROACHES

Neither of the above measurement options permit the direct determination of chemical (or, where appropriate, biological) species for the particles in question. Because none of the research so far relating to health effects has ruled out the role of species, other options must be kept open which might retain this possibility. For small particles selected and classified by either of the two preceding approaches, it is unlikely that sufficient material could be recovered (e.g., on a back-up filter) to enable determination by the techniques currently used routinely or aerosol quantitation (e.g., ICP-AES, X-ray diffraction, etc.). However there are exquisite analytical techniques now available which might be applicable. For example, Cohen *et al.* (1998) have recently developed a method to measure the size distribution and number concentration of acidic particles of size less than 200 nm based on the use of iron nano-films. Such detectors develop reaction sites when exposed to acid particles, while exposure to non-acidic particles results in no detectable surface deformations. Films are examined by means of *scanning force microscopy* (SFM) for the enumeration of reaction sites, and this allows the examination of particles in the size range of interest by high-quality, three-dimensional imaging. This method is superior to scanning or transmission electron microscopy (SEM and TEM respectively) because it is relatively inexpensive, does not require sample treatment, and can be performed at ambient pressure. Also, the depth of field is superior to that of TEM and its resolution is superior to that of SEM. Such techniques provide exciting opportunities for future research and development.

12.4 Conclusions

That there is at present no health-related particle size-selective criterion for very, very small particles reflects the fact that there is insufficient knowledge about cause and effect to permit such a criterion. But, because of the interest stimulated by concerns about health effects in populations exposed to the ambient environment, as well as questions about some workplaces, it is likely that extensive new research in the years ahead will address the problem. Nevertheless, we may for the time being speculate — with some confidence — that a future standard for very, very small particles will be expressed in terms of the number of particles less than a certain diameter. That diameter may be in the range from a few nm up to a few 10s of nm. When that criterion eventually emerges, it is clear that (at least) some of the required instrumentation will be available.

References

Agarwal, J.K. and Sem, G.J. (1980), Continuous flow, single particle-counting condensation nucleus counter, *J. Aerosol Sci.*, 11, pp. 343-357.

Chen, D.R., Pui, D.Y.H., Hummes, D., Fissan, H. *et al.* (1998), Design and evaluation of a nanometer aerosol differential mobility analyzer (Nano-DMA), *J. Aerosol Sci.*, 29, pp. 497–509.

Chen, L.C., Wu, C.Y., Qu, Q.S. and Schlesinger, R.B. (1995), Number concentration and mass concentration as determinants of biological response to inhaled irritant particles, *Inhal. Toxicol.*, 7, pp. 577-588.

Cohen, B.S., Li, W., Xiong, J.Q. and Lippmann, M. (1998), Detecting H^+ in ultrafine ambient aerosol utilizing iron nano-film detectors and scanning force microscopy, *Appl. Occup. Environ. Hyg.*, in press.

Dockery, D.W., Pope, C.A., Xu, X. *et al.* (1993), An association between air pollution and mortality in six U.S. cities, *N. Eng. J. Med.*, 329, pp. 1753-1759.

Donaldson, K., Li, X.-Y. and MacNee, W. (1998), Ultrafine (nanometer) particle mediated lung injury, *J. Aerosol Sci.*, 28, pp. 553–560.

Kittelson, D.B. (1998), Engines and nanoparticles: a review, *J. Aerosol Sci.*, 29, pp. 575–588.

Knutson, E.O. and Whitby, K.T. (1975), Aerosol classification by electrical mobility: apparatus, theory and applications, *J. Aerosol Sci.*, 6, pp. 443-451.

Oberdörster, G., Ferin. J., Gelein, R. et al. (1992), Role of the alveolar macrophage in lung injury: studies with ultrafine particles, Environ. Health Persp., 97, pp. 193-199.

Oberdörster, G., Gelein, R., Ferin, J. *et al.* (1995), Association of particulate air pollution and acute mortality: involvement of ultrafine particles, *Inhal. Toxicol.*, pp. 111-124.

Pope, C.A., Thun, M.J., Namboodiri, M.M. *et al.* (1995), Particulate air pollution as a predictor of mortality in a prospective study of U.S. adults, *Am. J. Resp. Crit. Care Med.*, 151, pp. 669-674.

Preining, O. (1998), The physical nature of very, very small particles and its impact on their behavior, *J. Aerosol Sci.*, 29, pp. 481–495.

Pui, D.Y.H. (1996), Direct-reading instrumentation for workplace aerosol measurements — a review, *The Analyst*, 121, pp. 1215-1224.

Pui, D.Y.H., Brock, J.R. and Chen, D.-R. (Guest Editors) (1998), Nanoparticles: a new frontier for aerosol research, J. *Aerosol Sci.*, (Special Issue), 29 (5/6), pp. 600-1200.

Rendall, R.E.G., Phillips, J.I. and Renton, K.A. (1993), Death following exposure to fine particulate nickel from a metal arc process, *Ann. Occup. Hyg.*, 38, 9 pp. 21-930.

Seaton, A. (1996), Particles in the air: the enigma of urban air pollution, *J. Roy. Soc. Med.*, 89, pp. 604-607.

Seaton, A., MacNee, W., Donaldson, K. *et al.* (1995), Particulate air pollution and acute health effects, *The Lancet*, 345, pp. 176-178.

Stolzenburg, M.R. and McMurry, P.H. (1995), An ultrafine condensation nucleus counter, *Aerosol Sci. Tech.*, 14, pp. 48-65.

Vincent, J.H. (1995), *Aerosol Science for Industrial Hygienists*, Pergamon Press, Oxford, U.K.

Weber, R.J., Stolzenburg, M.R., Pandis, S.N. *et al.* (1998), Inversion of ultrafine condensation nucleus counter pulse height distributions to obtain nonparticle (3 to 10 nm) size distributions, *J. Aerosol Sci.*, 29, pp. 601–615.

◆ ◆ ◆ ◆ ◆

Chapter 13

PERFORMANCE ACCEPTANCE CONSIDERATIONS FOR WORKPLACE AEROSOL SAMPLERS

Göran Lidén

National Institute for Working Life, Solna, Sweden

13.1 Introduction

The American Conference of Governmental Industrial Hygienists (ACGIH) has promulgated particle size-selective (PSS) threshold limit values (TLVs) for inhalable particulate mass (IPM), thoracic particulate mass (TPM), and respirable particulate mass (RPM). These are presented in the forms of mathematical expressions for the desired sampling efficiency as a function of aerodynamic particle diameter (ACGIH, 1999). Any sampler which demonstrably follows a given sampling convention is considered a valid sampler for that convention, and it does not need to be a specific sampling device. This therefore represents a major difference with sampling aerosol using a specific sampler — for example, the closed-face 37-mm filter cassette (for 'total' aerosol) or 10-mm nylon cyclone (for respirable aerosol) — at a specified flowrate. These samplers, and others like them, are often considered in themselves as "implied" standards. In principle an implied standard sampler, by its choice and use, always samples correctly with respect to that standard (although conditions may be known in practice where it does not operate in a trustworthy way). This sampler in turn sets the reference against which other, alternative samplers are compared or evaluated.

The specification of criteria such as IPM, TPM and RPM is different in that it requires that the sampling convention be external to the sampler. In order to determine whether a particular sampler follows the sampling convention or not, it is required that the sampler be tested and its performance relative to the sampling convention be assessed within a rational framework. This leads to a type of performance test that is qualitatively different from a test of an implied standard sampler, which is mainly tested against itself for different environmental and other conditions (McCawley, 1985; Kenny and Lidén, 1989, John, 1993; Lidén, 1994).

The sampling conventions are very compact and simply formulated. On the other hand, to verify whether a given sampler actually follows the sampling convention requires extensive testing. In Europe, the standard for the sampling conventions themselves (corresponding to the ACGIH definitions described in this book) required only 9 typed pages (Comité Européen de Normalisation (CEN), 1993), whereas the final draft standard for the performance test required as many as 34 typed pages.

During the 1970s, the term "*sampler performance*" was a synonym for the sampler's penetration (or, conversely) efficiency curve; however, since the 1980s it has come to mean a statistical determination of "how well" (in some meaning) the sampler matches the intended sampling convention. In the earlier use of the term, sampler performance could be determined independently of any sampling convention, whereas according to current usage, the term is bound to a sampling convention. In this chapter, the approach adopted is that "performance" relates to a specific sampling convention. Not only is the actual sampling efficiency considered, but consideration is also given to other factors which might influence the sampled — and subsequently analyzed — particles. In this context, a "performance test" must not be confused with all the experiments/tests that are done by scientists or engineers during the development of a sampler. It is something that is done with the final version of the sampler once the optimal design flowrate and the intended position where to wear the sampler are decided.

The issues that are raised in this chapter are drawn from the body of work and deliberations that have been carried out within the European Community, acknowledging the importance the members of that community are giving to the matter of health-related aerosol sampling. It is important to note that the European Community is not a federal state (like, for example, the United States). Rather, it is a community of independent states which have passed some of their legislative power to the Council of Ministers, the European Commission, and the European Parliament respectively. Concerning hazardous chemical and biological substances, a Council Directive has been issued entitled "*Protection of workers from the risks related to exposure to chemical, physical and biological agents at work, 88/642/EEC.*" It is presently under revision. All directives must be incorporated into the laws of each member state. Limit values, however, are set by each government. In order to harmonize the strategy for assessment of exposure and the performance criteria for sampling equipment and analytical methods, the CEN (1999) is working on drafting a range of standards (through Technical Committee 137). These standards are general and oriented towards the performance of the equipment/methods to be used and are not specific for a certain analysis. Within the European context, there are several reasons for focusing the standardization work on such performance-based criteria. But, most importantly, it helps overcome the 'political' difficulties in coming to agreement on any particular sampling method.

13.2 IMPLIED IDEALIZED MODEL BEHIND A PERFORMANCE TEST

Implied in any test of sampler performance is a model of the reality in which the sampler is to be used. It may, for example, be assumed that the sampler test does, or should, incorporate (all) possible situations encountered during sampling at different workplaces. That is, one should not encounter any surprises when using the sampler at any workplace. On the other hand, the test situation may be idealized, and subsequently, when the sampler is used in workplace sampling, instances may occur where the sampler deviates from what can be deduced from the test. Two examples of simple and idealized test situations are, respectively, studies of inhalable dust (tested in a wind tunnel) and sampling in calm air (tested in a static dust chamber). In the first example, the dust source is far from the exposed worker, and the airborne particles have lost all memory of the aerosol source. All particles are carried to the worker with the wind. Both the windspeed and the particle concentration are constant over the cross-section of the wind tunnel. The worker shows no preferred direction to the dust source/wind, and there are neither obstacles disturbing the air movement nor local exhausts (as there was no local source). There is a lower limit for the windspeed below which it becomes experimentally difficult to establish constant aerosol concentrations within the approaching air. Thus, such wind tunnel experiments are very difficult to perform at the low windspeeds which pertain to many workplaces. In the second example, the dust source is above the worker with a constant concentration across the horizontal plane. The dust is sedimenting under the influence of gravity, with the air movement caused only by the samplers, or possibly also by breathing (in the case of a human subject or a mannequin). There is no turbulence to transport dust in and out of the sampled air flow. In contrast to both these examples, the air in workplaces is neither calm nor is its speed very high (indoors it is usually in the range 0.1–0.3 m/s), except when close to forced ventilation or open doors. The worker is working with or handling the dust source, which generally is located in front of him/her, often lower than his/her face. The source may be a point source, or an area source, and possibly located on a larger object which disturbs the air flow. The particles may be ejected with high momentum or in concentrated air jets which would enable them to travel by themselves towards the worker without the help of external air movements or suction from inhaled or sampled air flows. Local exhaust may also be present.

The ideal performance test should be simple, quick, cheap, and valid with respect to all possible workplaces. The last requirement is presumably impossible, and therefore a real performance test must be based on an idealized model, for example, one of those described above. However, even such a test would be neither simple, quick nor cheap, and samplers with similar results in such an idealized test may differ significantly at workplaces where the assumptions of the test are violated. For example workplace aerosols and the selected test aerosol may exhibit different bounce capacities when they come into contact with surfaces (Hall *et al.*, 1990) and different susceptibility to de-agglomeration by aerodynamic shear forces within the sampler. Therefore, any performance test is a compromise between opposing objectives, and the theoretical model behind the agreed test, as well as the consequences of its deviations from real workplaces, must be accepted but not forgotten. A performance test may either be very formally structured (i.e., strictly prescribing test geometry, test aerosol, measuring system, number of tested candidate samplers, what to test, etc.) or informally prescribed (i.e., most is left to the testing laboratory to decide, but with requirements on the quality of measurement of all independently varied or measured, dependent, variables). The first test type would enable different laboratories to do identical performance tests (which should, in principle, give identical results), but there is no guarantee that different types of samplers which pass the performance test would obtain similar concentrations in workplace sampling at different conditions. The second test type invites the test laboratory to design an optimal test depending on how the candidate sampler works and its intended use. With this latter type of test, one may expect the test to be better tailored to the specific requirements of the candidate sampler, but there would be no guarantee that different test laboratories would obtain identical results.

13.3 Implied Standard Sampler vs. Samplers for Sampling Conventions

Samplers developed in the 1960s and 1970s, and presently being used to sample respirable dust and so-called total dust, are often in themselves implied standards. In U.S. coal mines, for example, a specific cyclone is required for sampling respirable coal dust; that is, the 10-mm nylon cyclone is stipulated in U.S. law (*Code of U.S. Federal Regulations,* 1982). In British coal mines, coal dust is measured in the return air with a stationary horizontal elutriator, which in calm air perfectly follows the British Medical Research Council (BMRC) convention for respirable dust (Lidén and Kenny, 1991), and where, as mentioned in Chapter 8, the latter sampling convention was itself designed as the theoretical penetration of a horizontal elutriator. At present, only cyclones for respirable dust (excluding British and U.S. coal mines) and samplers for inhalable dust used in some countries in Europe (e.g., HSE, 1997) can be said to be designed to follow a specific sampling convention. 'Total' aerosol, on the other hand, is measured worldwide, but many countries have their own definition of total dust. As related in Chapters 3 and 6, the definition of 'total' dust is often what is sampled by a sampler generally used in that country (often an open or closed-face filter cassette), irrespectively of how that aerosol fraction relates to the total airborne concentration or how sensitive it is to variations in ambient conditions, such as windspeed. In most countries, there is a virtual one-design monopoly governing which 'total' aerosol sampler is used; thus, the sampler commonly used may be viewed as an implied standard in itself (e.g., as is the case for the closed-face 37-mm plastic cassette in the U.S.).

An interest which has surfaced in the U.S. and Europe since the end of the 1980s is the international harmonization of dust sampling. It would probably not be possible to promote any specific national reference sampler as the "world reference." But by requiring samplers to follow a sampling convention, rather than being made according to a specific design, many different, originally national, samplers may be found to be suitable. With this in mind, the CEN has produced standards with the aim to ensure that different samplers can be used, provided their performance meets certain requirements (CEN, 1993, 1994 and 1999). In order to better follow the new international sampling conventions, nationally used samplers (standards in themselves) may have to be optimized (by varying the flowrate or some other easily changed dimension) (Maynard, 1993; Lidén and Gudmundsson, 1996) if they deviate too much from the sampling convention in question. The CEN approach will stimulate the development of new — and hopefully better — samplers, as compared with the previous system in which all new samplers were (implicitly, at least) required to agree with the national standard instrument, whatever the sampling efficiency of that standard instrument.

13.4 METHODS FOR COMPARING SAMPLING EFFICIENCY WITH A SAMPLING CONVENTION

How well a sampler emulates a prescribed sampling convention can be expressed in two different ways. The degree of conformity may either be expressed as a deviation in sampling efficiency (as a function of particle size) or as a deviation in sampled concentration (an aggregated value over the relevant particle size range). The latter is of more use to the industrial hygiene community, whereas the former is of more use to designers of aerosol samplers and aerosol scientists. To determine if a candidate sampler follows a sampling convention "sufficiently well," one must first define both a test method and a set of rules by which the sampler may be judged. The test is based on a mixture of science (aerosol science and statistics), politics (a decision about the meaning of "sufficiently well"), and economics (what it may cost to do a test). All three develop and change over time.

In principle, three different methods exist to establish whether a new (candidate) sampler follows a given sampling convention. The first, and simplest, method is possible only if a reference sampler exists which perfectly follows the sampling convention. In such a case, the mass concentrations sampled simultaneously with the candidate and the reference sampler may be compared directly, preferably in a laboratory, but in some cases possibly also in the field. Such a comparison must be repeated for several aerosols with different size distributions. So far this has only been possible with the BMRC sampling convention for respirable dust because the horizontal elutriator provides the desired 'absolute' reference. But for all other sampling conventions, no true reference sampler exists. Both the experiment and the statistical evaluation of a comparison between a candidate sampler and reference sampler are fairly simple and straightforward. A possible requirement is that for each size distribution, both the bias and variance of the population of the candidate sampler should lie within some specified limits (Tomb *et al.*, 1973; Caplan *et al.*, 1977a and b). This method may also be used if a good validated sampler for the sampling convention already exists and can be used as a reference sampler (under the assumption that the reference sampler itself would first have been tested and validated using any of the two methods described below).

In the second method, no reference instrument that theoretically follows the sampling convention exactly exists against which the candidate sampler can be compared. Therefore, the sampling efficiency of the candidate sampler must be determined experimentally for a range of particle sizes, so that its characteristic sampling efficiency curve may be constructed, usually by non-linear regression. The resultant sampling efficiency curve of the candidate sampler is then qualitatively compared with the prescribed sampling convention. Alternatively, the individual sampling efficiency data for the candidate sampler may be plotted on a diagram together with the sampling convention, and acceptance bands about the conventional curve may be employed to help identify the extent to which the measured sampling efficiency deviates from the convention. Requirements for the 'cut-off' particle size and the slope of the candidate sampler's sampling efficiency curve are also considered in this approach (McCawley, 1985; Deutches Institute fur Normung, 1988).

The third method is more quantitative. Here, the candidate sampler's efficiency data are again used to construct a penetration curve. But in this case, the resultant curve is used to compute the fraction of the airborne mass concentration that is collected by the collecting substrate (usually a filter) of the candidate sampler, for a set of log-normal size distributions, and these are compared with the corresponding masses that would be collected by a hypothetical ideal sampler which perfectly follows the sampling convention. The relative difference in concentration is termed the "*bias*." This bias is calculated for a set of particle size distributions covering the range of relevant interest, and the performance of the candidate sampler is determined from the bias over the set of size distributions (Caplan *et al.*, 1977a and b; Bartley and Breuer, 1982). In this case, the mass bias, determined over the whole set of relevant particle size distributions, should be below a specified value. In an advanced version of this test, both the variation between the individual samplers and the experimental uncertainty may be incorporated into the statistical evaluation (Bowman *et al.*, 1984; Kenny and Lidén, 1989; Bartley *et al.*, 1994).

All three methods have their advantages and disadvantages. The first method is relatively inexpensive and easy to use, but requires several different test aerosols with different size distribution characteristics. The method can only be used when a reference instrument exists, e.g., the horizontal elutriator for the BMRC convention in calm air. Differences obtained might be difficult to analyze, but they are expressed in a way that is easily understood by the end user. The second method gives more information, but the relationship between the acceptance bands and measured mass concentrations is not always obvious (Bartley and Doemeny, 1986). The third method gives results that are more directly related to the needs of the end users. However, the validity of the calculations of sampled concentrations in the real world have been questioned in some quarters. In one case, for example, an optimization of the nylon cyclone (with changed flowrate) based on predicted computed fractions, could not — in experiments in a coal mine — be shown to be better than the standard nylon cyclone at the ordinary flowrate (Treaftis *et al.*, 1984).

13.5 PARTS OF A PERFORMANCE TEST

Many performance tests of aerosol samplers have been carried out during the last 25 years, and the results are in many cases published, either as peer-reviewed papers or as publicly available reports; among these are Bowman *et al.*, 1984; Lidén and Kenny, 1991; Bartley *et al.*, 1993; Lidén and Gudmundsson, 1996. These reports differ widely in the scope of the performance test. Standards or recommendations for performance tests have been published by several organizations (Gundersen *et al.*, 1980; McCawley, 1985; EPA, 1987; Deutsches Institut für Normung, 1988; Kennedy *et al.*, 1995; CEN, 1999). These tests will not be reviewed here. Instead a general outline of a performance test will be presented, mainly drawing from the draft standard of the CEN technical committee. In several aspects the CEN draft standard is more similar to the EPA specification of the PM_{10} sampler than to the 10-mm nylon cyclone specified during the 1970s and 1980s by the National Institute for Occupational Safety and Health (NIOSH), Occupational Safety and Health Administration (OSHA), and the Mine Safety and Health Administration (MSHA). Presently, however, EPA is moving away from a concept based on a performance standard (which was originally proposed for the PM_{10} sampler) towards a concept of a reference sampler based on a specific design specified by a drawing (as for $PM_{2.5}$) (EPA, 1997).

A performance test of a candidate sampler will generally include five subparts. The first is a theoretical analysis, or critical review (CEN, 1999) of all stages in the use of the candidate sampler. The subsequent stages are the actual laboratory tests, the analysis (statistical and otherwise) of these results, the conclusion and recommendations, and finally a handling (or usability) test often combined with a workplace sampling exercise with the candidate sampler.

CRITICAL REVIEW

The critical review consists of a theoretical review of all aspects of the aspiration, transport, separation and collection, retention and/or retrieval of the sample, and how the use and handling of the sampler may interfere with the intended sampling. This includes the cleaning and preparation of the sampler at the laboratory, its use at a workplace, shipment back to the laboratory, and the subsequent storage and analysis. Examples of factors to consider are: integrity of collection media and sample during sampling and transport; intervals for cleaning the sampler and replacement of collection media; air flow stability and pulsation; possibility of non-actively sampled particles becoming part of the sample; sensitivity to particle bounce, physical phase (solid, liquid or gas), particle and sampler electrical charge; the shape of sampled particles; sensitivity to ambient temperature, pressure, humidity and air movement (speed, direction and turbulence); sampler position on the worker and variability among sampler individuals; and whether the sampler is so impractical or complicated to wear and use that in practical workplace sampling it is probable it may not be used as intended or conceived. A recent review of sampler design indicates some details of current and new samplers that should receive critical evaluation (Baron, 1998).

The critical review should identify the critical aspects of the use of the sampler and from this determine what needs to be known about the sampler in order to be able to judge its performance. Some critical aspects of the candidate sampler may be so well known for samplers of this type that they do not need to be tested, whereas other critical aspects must be tested in order to be able to determine the performance of the candidate sampler. The critical analysis of different types of samplers may result in requirements for different tests.

LABORATORY TEST

Based on the critical review, it is decided what should be tested and how it should be carried out. One of the major decisions is whether it will be necessary to determine the complete sampling efficiency curve, or it would be sufficient to only measure an integrated value (i.e., the sampled concentration). Unless there is evidence (e.g., from current or future research) to the contrary, samplers intended for use as personal samplers should be tested while mounted on a mannequin. It is not necessary that the mannequin be breathing in and out during the experiment as the external air velocities and patterns caused by a breathing mannequin seem to be different from those caused by an actual breathing human (Bradley *et al.*, 1994). If the internal losses to the sampler walls between the sampler inlet plane and the actual separation stage are negligible for the thoracic and respirable fractions, the sampling efficiency for samplers for these fractions may be determined in two steps. In the first step, the candidate sampler's aspiration efficiency is determined in a wind tunnel if moderate or high windspeeds (e.g., 0.5 m/s or above) may be encountered at the use of the candidate sampler. In the second step, the internal penetration or separation efficiency is determined in a much simpler experimental set-up in a static dust chamber or by delivering the test aerosol directly to the separation stage.

In some cases, it may be possible to test the performance of the candidate sampler by a comparison with a validated reference sampler using polydisperse test aerosols. In this case three or more test aerosols with different particle mass median aerodynamic diameters and geometric standard deviations would be necessary. These should span the relevant range of the fraction to be sampled, and the ratio of the fraction sampled to the total airborne particle concentration should be neither too high nor too low.

In the laboratory experiment, the concentration measured by the candidate sampler is compared with the total airborne particle concentration in the test system. Polydisperse test aerosols may be used if either a measuring system capable of counting particles according to their aerodynamic size is available or it is considered not necessary to determine the sampling efficiency as a function of particle size. If calibration curves for particle aerodynamic size or correction factors for particle density and shape are used, their uncertainty should be small ($\approx 5\%$), and the resulting uncertainty of the measured particle sizes should be evaluated. Monodisperse test aerosols used for testing samplers with a sharper efficiency curve need to have a narrower distribution than those used for testing samplers with a more shallow efficiency curve. The particle size range of polydisperse test aerosols used with a particle counter, should — in order to obtain good counting statistics — extend well over the expected particle size range relevant to the candidate sampler. The size of the wind tunnel or static dust chamber used in the test should be much larger than the candidate sampler and the sampler used to measure the total airborne particle concentration. For small, candidate personal samplers, several individual items should be tested. If a particle counter is used in the test, the candidate sampler individuals should be tested sequentially, with several runs per individual to obtain more than one measured sampling efficiency value. When a particle counter is not used, several candidate samplers and samplers which sample the total airborne concentration may be placed in the test system simultaneously. It is important that both the concentration sampled by the candidate sampler and the total airborne concentration are measured accurately and precisely. If the total airborne concentration varies over time or space, it is problematic to estimate it at the location and time that the candidate sampler samples the test aerosol. (e.g., see Chalmers, 1995) To determine the total airborne concentration in a wind tunnel is straightforward (using thin-walled, sharp-edged probes operating isokinetically), whereas it is more problematic in a static dust chamber. In the latter case, one could use any method with a sampling efficiency approximately equal to unity for all particle sizes of interest. The analytical determination of the sample must be adequate to assure that the sampling efficiency is determined with only a few percent uncertainty.

If the critical analysis does not show that strong interactions can be expected between some factors, the different factors may be treated as independent of each other, and hence tested independently.

When the experimental plan is designed, it would be wise to consult a statistician to ensure that the experimental design is efficient both from the point of view of bias and uncertainty. The total experimental uncertainty of the performance test should be evaluated.

CALCULATIONS AND STATISTICAL ASSESSMENT

As the samplers are intended to measure the aerosol concentration, it is advisable that all measured data on sampling efficiency as a function of particle size (and other parameters) be converted into sampled aerosol concentrations (or fractional concentrations) so that the bias and precision in the measured concentrations may be estimated. This requires that sampling efficiency curve data are first reduced to a mathematical function. The simplest way to do this is to estimate the sampling efficiency function with a polygon whose value is constant and equal to the measured sampling efficiency for each particle size interval (Kenny, 1995). (Note: This method would give large errors if the sampling efficiency curve is a steep function of particle aerodynamic size.) Other methods to obtain a sampling efficiency function from the data include (transformed) linear and non-linear regression and spline curves. There is no obvious single recommended model to use for fitting the data (e.g., a model similar to the sampling convention in question). However, all models which fit data without (or with only marginal) lack of fit (Draper and Smith, 1981) would give similar results when used in further calculations. When the curve is obtained, the sampled fractional concentration is calculated for a set of log-normal particle size distributions. The bias relative to the sampling convention would be a function of the mass median aerodynamic diameter and geometric standard deviation of the particle size distribution, and this can easily be plotted as a contour diagram (called a "*bias diagram*"). Each level for each environmental or other factor for which the sampling efficiency is determined as a function of particle size will generate one diagram for its bias.

The object of the statistical evaluation of the laboratory experiment is to obtain data that make it possible to statistically judge the performance of any individual sampler belonging to the whole population of the candidate sampler individuals and not only the actually tested individuals. Therefore, the conclusion of the performance test is stated as the fraction of the concentrations measured with individuals of the candidate sampler that, with a specified confidence (as not all individuals from the population were tested), will deviate less than a given amount (maximum difference) from that which would have been obtained using an ideal hypothetical sampler perfectly following the sampling convention. The three resulting parameters (fraction, confidence, and maximum difference) are interrelated. If any two are specified, the third expresses the result and depends on the bias and precision of the candidate sampler. A diagram similar to the bias map may be drawn which shows the dependence on particle size distribution (mass median aerodynamic diameter and geometric standard deviation for hypothetical uni-modal log-normal size distributions). The result may therefore be expressed as either: (a) the fraction of the sampler individuals which, with a given confidence, deviates less than a given amount from the sampling convention; (b) the confidence that a given fraction of the sampler individuals deviates less than a given value from the sampling convention; or (c) the maximum difference that a given fraction of the sampler individuals — with a given confidence — deviates from the sampling convention. NIOSH has published extensive reports on this subject (Bartley and Fischbach, 1993; Kennedy *et al.*, 1995), and it is also fundamental to the CEN draft standard on aerosol sampler performance tests (CEN, 1999). The latest evaluation of the 10-mm nylon cyclone was reported as a performance test (Bartley *et al.*, 1994), and Kenny and Bartley (1995) have statistically analyzed much of the data for several PM_{10} samplers. A European wind tunnel experiment with several samplers for the inhalable fraction or 'total' aerosol has also been evaluated in this manner (Kenny *et al.*, 1997).

13.6 CONCLUSIONS AND RECOMMENDATIONS

A performance test of a sampler contains so much information that it would be unwise to leave it to every end-user of the sampler to try to interpret themselves whether to use that sampler or not. Instead it is advisable that a performance test contains requirements that enable a judgment to be passed on the sampler. The possible judgment shall only be one of two, or possibly three, alternatives: accepted; rejected; or uncertain.

The test parameter to be used in the judgment may be either the bias relative to the sampling convention or the fraction of the population of samplers that with a certain confidence deviates less than a certain amount from the sampling convention. The former value, the bias, is more easily understood intuitively, but the latter is a better test parameter as it neither assumes that all samplers give identical results, nor that the laboratory experiment was extensive enough for no uncertainty to be associated with, for example, the bias. For example the test parameter could be selected as the fraction of samplers which, with 90% confidence, will be within ±25% of a sampler that perfectly follows the sampling convention: in short, the fraction of samplers that will be within ±25% of the sampling convention with 90% confidence.

Each level for each environmental or other factor for which the sampling efficiency is determined as a function of particle size will generate a test parameter of how closely a fraction of the sampler population follows the sampling convention. This fraction of the sampler population will in itself be a function of the aerosol mass median aerodynamic diameter and geometric standard deviation. It is therefore important that the size distributions used for the evaluation be properly selected. They must span the distributions likely to be encountered at workplaces. When the sampled fraction is very low, its value depends either on the tail of the size distribution or on the tail of the sampling efficiency curve. Neither of these tails would be determined as accurately as the main aspects of the two curves, and size distributions with low sampled fractions should therefore be excluded from the calculations of the test.

For those aspects of the performance test that are based on determining the sampling efficiency as functions of particle size (subsequently converted into sampled concentrations), there are two possible requirements: either all, or only a fraction, of the test parameters for the size distributions of interest shall be above or below a critical value. For samplers that are generally used today, and which have not been designed to provide a good emulation of the sampling convention, a requirement that the fraction of the sampler population shall be acceptable for all size distributions of interest is probably so strict that none would pass.

The acceptable conditions for use of the sampler would then be determined by the valúes of the levels of experimentally or otherwise tested factors for which the test parameter — the fraction of the sampler population — is above a critical value. As stated above, all factors may be treated as independent of each other, and therefore each may be evaluated independently of the other. However, it would be unwise if, for example, a sampler would be acceptable at a windspeed of 2.5 m/s, but not at lower windspeeds; for moderately electrically charged particles, but not for particles in Boltzmann equilibrium; for very bouncy solid particles, but not for liquid or soft solid particles (or the opposite); and so on. For factors like these, the acceptance region must include the most simple level and extend outwards towards a limit at which the sampler is no longer accepted. For a sampler that passes the test, the test report must be publicly available.

Handling and usability are important matters for the practical user of aerosol samplers. But for these there is no appropriate or specific test in the sense that its outcome would lead to the acceptance or rejection of the candidate sampler. It is not necessary that any such evaluation is carried out by the same laboratory which performed the quantitative performance test. But a field test **is** necessary before a candidate sampler can be recommended for use. The object here is to verify that, in real-world workplaces, the candidate sampler may be used as intended and does not interfere with the worker. For example, it should be easy to attach and position the candidate sampler on the worker and detach it from him/her and also possible to verify the flowrate through it when worn by a worker, without disturbing him/her too much. But, if the candidate sampler is difficult to operate or to handle in the laboratory, short-cuts will be taken in workplaces because of time stress.

It would also be valuable if the candidate sampler were to be compared with an accepted reference sampler during workplace sampling, in order to verify that similar concentrations are sampled also at real workplaces as opposed to test laboratories. The other sampler may be a more complicated sampler used for scientific investigations and so not necessarily one intended for routine sampling.

It is important that any sampler to be tested is marked with the name of the manufacturer and the type number. The manual should at least contain the following: which sampling conventions the sampler follows (and at which sampling flowrate); known limitations to the use of the sampler and problems that may be encountered; details of the particle collection substrate to use; how to cleanse, maintain, set up, and adjust the sampler; requirements for an external pump; recommended batteries and battery charger, as well operating time, if applicable.

How to accurately determine the sampling flowrate and weighing procedures for filters, etc. are not covered in here as those problems are not specific to the performance of a candidate aerosol sampler.

Finally, actual workplace comparisons between a reference sampler (generally following the sampling convention) and a candidate sampler which might be intended to become an equivalent sampler at that specific workplace are not covered in this chapter. But suffice it to say that the correlation between the two samplers must be good if the random variation shall not destroy any useable systematic relation for converting concentrations between the two samplers. Both the systematic difference between the two samplers and their covariance will depend on the processes that generate dust at that workplace, and it will therefore be a strong function of the workplace itself (and its dust generating processes), and not only be a function of the physical operation of the two samplers.

REFERENCES

American Conference of Governmental Industrial Hygienists (ACGIH) (1999), *Threshold Limit Values for Chemical Substances and Physical Agents and Biological Exposure Indices*, American Conference of Governmental Industrial Hygienists, Cincinnati, OH, 1999.

Baron, P.A. (1998), Personal aerosol sampler design: a review, *Appl. Occup. Environ. Hyg.*, 13, pp. 313-320.

Bartley, D.L. and Breuer, G.M. (1982), Analysis and optimization of the performance of the 10-mm cyclone, *Am. Ind. Hyg. Ass. J.*, 43, pp. 520-528.

Bartley, D.L. and Doemeny, L.J. (1986), Critique of 1985 ACGIH Report on Particle Size-Selective Sampling in the Workplace, *Am. Ind. Hyg. Ass. J.*, 47, pp. 443-447 and A498-A500.

Bartley, D.L. and Fischbach, T.J. (1993), Alternative approaches for analyzing sampling and analytical methods, *Appl. Occup. Environ. Hyg.*, 8, pp. 381-385.

Bartley, D.L., Chen, C.-C., Song, R. *et al.* (1994), Respirable aerosol sampler performance testing, *Am. Ind. Hyg. Ass. J.*, 55, pp. 1036-1046.

Bowman, J.D., Bartley, D.L., Breuer, G.M. *et al.* (1984), Accuracy criteria recommended for the certification of gravimetric coal mine dust samplers, NTIS No. 85222446, National Institute of Occupational Safety and Health, Cincinnati, OH.

Bradley, D.R., Johnson, A.E., Kenny, L.C. *et al.* (1994), The use of a mannikin for testing personal aerosol samplers, *J. Aerosol Sci.*, 25, pp. 155-156.

Caplan, K.J., Doemeny, L.J. and Sorenson, S.D. (1977a), Performance characteristics of the 10-mm cyclone respirable mass sampler: Part I — Monodisperse studies, *Am. Ind. Hyg. Ass. J.*, 38, pp. 83-95.

Caplan, K.J., Doemeny, L.J. and Sorenson, S.D. (1977b), Performance characteristics of the 10-mm cyclone respirable mass sampler: Part II — Coal dust studies, *Am. Ind. Hyg. Ass. J.*, 38, pp. 162-173.

Chalmers, C. (1995), *Calculating Efficiencies*, DataStat Consultants, London, UK.

Comité Européen de Normalisation (CEN) (1993), *Workplace Atmospheres – Size Fraction Definitions for Measurement of Airborne Particles*, EN481, CEN, Brussels, Belgium.

Comité Européen de Normalisation (CEN) (1994), *Workplace Atmospheres – General Requirements for the Performance of Procedures for the Measurement of Chemical Agents*, EN482, CEN, Brussels, Belgium.

Comité Européen de Normalisation (CEN) (1999), *Workplace Atmospheres – Assessment of Performance of Instruments for Measurement of Airborne Particle Concentrations*, TC137/WG3/N192, CEN, Brussels, Belgium.

Deutsches Institut für Normung (1988), Luftbeschaffenheit am arbeitplatz – einatembarer und alveolengangiger staub [Workplace air quality – inhalable and respirable dust], NLuft/AA 11 AK No. 8-88, Deutsches Institut für Normung e.v., Berlin, Germany

Draper, N. and Smith, H. (1981), *Applied Regression Analysis, 2nd Edition*, Wiley Series in Probability and Mathematical Statistics (W.A. Shewhart and S.S. Wilks, Eds.), John Wiley and Sons, New York.

Gundersen, E., Anderson, C., Smith, R.H. *et al.* (1980), Development and validation of methods for sampling and analysis of workplace toxic substances, Report No. 80-133, National Institute for Occupational Safety and Health, Cincinnati, OH.

Hall, D.L., Upton, S.L., Marsland, G.W. *et al.* (1990), Wind tunnel measurement of the particle collection efficiency of the Institute of Occupational Medicine ambient inhalable and thoracic particle sampling heads, Report No. LR806, Warren Spring Laboratory, Stevenage, UK.

Health and Safety Executive (HSE) (1997), *General Methods for Sampling and Gravimetric Analysis of Respirable and Total Inhalable Dust*, MDHS14, Health and Safety Executive, London, UK..

John, W. (1993), Instrument performance and standards for sampling of aerosols, *Appl. Occup. Environ. Hyg.*, 8, pp. 251-259.

Kennedy, E.R., Fishbach, T.J., Song, R. *et al.* (1995), Guidelines for air sampling and analytical method development and evaluation, Report No. 95-117, National Institute of Occupational Safety and Health, Cincinnati, OH, 1995.

Kenny, L.C. (1995), Pilot study of CEN protocols for the performance of workplace aerosol sampling instruments, Report IR/L/DS/95/18, Health and Safety Laboratory, Sheffield, U.K.

Kenny, L.C. and Lidén, G. (1989), The application of performance standards to personal airborne dust samplers, *Ann. Occup. Hyg.*, 33, pp. 289-300

Kenny, L.C. and Bartley, D.L. (1995), The performance evaluation of aerosol samplers tested with monodisperse aerosols, *J. Aerosol Sci.*, 26, pp. 109-126.

Kenny, L.C., Aitken, R.J., Chalmers, C. *et al.* (1997), A collaborative European study of personal inhalable aerosol sampler performance. *Ann. Occup. Hyg.*, 41, pp. 135-153.

Lidén, G. (1994), Performance parameters for assessing the acceptability of aerosol sampling equipment, *The Analyst*, 119, pp. 27-33.

Lidén, G. and Kenny, L.C. (1991), Comparison of measured respirable dust sampler penetration curves with sampling conventions, *Ann. Occup. Hyg.*, 35, pp. 485-504.

Lidén, G. and Gudmundsson, A. (1996), Optimisation of a cyclone to the 1993 international sampling convention for respirable dust, *Appl. Occup. Environ. Hyg.*, 11, pp. 1398-1408.

Maynard, A.D. (1993), Respirable dust sampler characterisation: efficiency curve reproducibility, *J. Aerosol Sci.*, 24, pp. S457-S458.

McCawley, M.A. (1985), Performance considerations for size-selective samplers, In: *Particle Size-Selective Sampling in the Workplace* (R.F. Phalen, Ed.), Report of the ACGIH Technical Committee on Air Sampling Procedures, American Conference of Governmental Industrial Hygienists, Cincinnati, OH, pp. 77-80.

Tomb, T.F., Treaftis, H.N., Mundell, R.L. *et al.* (1973), Comparison of respirable dust concentrations measured with MRE and modified personal gravimetric sampling equipment, Report No. RI7772, Bureau of Mines, Pittsburgh, PA.

Treaftis, H.N,, Gero, A.J., Kacsmar, P. *et al.* (1984), Comparison of mass concentrations determined with personal respirable coal mine dust samplers operating at 1.2 litres per minute and the Casella 113A gravimetric sampler (MRE), *Am. Ind. Hyg. Ass. J.*, 45, pp. 826-832.

Tsai, C.-J. and Shih, T.-S. (1995), Particle collection efficiency of two personal respirable dust samplers, *Am. Ind. Hyg. Ass. J.*, 56, pp. 911-918.

United States Department of Labor, Mine Safety and Health Administration, (1982), 30 Code of US Federal Regulation, Part 74.

United States Environmental Protection Agency (EPA) (1987), *Ambient air monitoring reference and equivalent methods*, 40 Code of US Federal Regulation, part 53.

United States Environmental Protection Agency (EPA) (1997), 40 Code of US Federal Regulation, Part 50, Chapter 7.